ATOM PROBE TOMOGRAPHY

Analysis at the Atomic Level

ATOM PROBE TOMOGRAPHY

Analysis at the Atomic Level

M. K. Miller
Oak Ridge National Laboratory
Oak Ridge, Tennessee

Springer Science+Business Media, LLC

Library of Congress Cataloging-in-Publication Data

Miller, M. K. (Michael Kenneth)
Atom probe tomography : analysis at the atomic level / by M.K. Miller.
p. cm.
Includes bibliographical references and index.
ISBN 978-1-4613-6921-9 ISBN 978-1-4615-4281-0 (eBook)
DOI 10.1007/978-1-4615-4281-0
1. Atom-probe field ion microscopy. 2. Tomography. I. Title.

QH212.A76 M56 2000
502'.8'2--dc21

00-034894

ISBN 978-1-4613-6921-9

Originally published by Kluwer Academic / Plenum Publishers, New York in 2000

10 9 8 7 6 5 4 3 2 1

A C.I.P. record for this book is available from the Library of Congress

"… in reality nothing exists but atoms and the void."

Democritus

Preface

The microanalytical technique of atom probe tomography (APT) permits the spatial coordinates and elemental identities of the individual atoms within a small volume to be determined with near atomic resolution. Therefore, atom probe tomography provides a technique for acquiring atomic resolution three-dimensional images of the solute distribution within the microstructures of materials.

This monograph is designed to provide researchers and students the necessary information to plan and experimentally conduct an atom probe tomography experiment. The techniques required to visualize and to analyze the resulting three-dimensional data are also described.

The monograph is organized into chapters each covering a specific aspect of the technique. The development of this powerful microanalytical technique from the origins of field ion microscopy in 1951, through the first three-dimensional atom probe prototype built in 1986 to today's commercial state-of-the-art three-dimensional atom probe is documented in chapter 1. A general introduction to atom probe tomography is also presented in chapter 1. The various methods to fabricate suitable needle-shaped specimens are presented in chapter 2. The procedure to form field ion images of the needle-shaped specimen is described in chapter 3. In addition, the appearance of microstructural features and the information that may be estimated from field ion microscopy are summarized. A brief account of the theoretical basis for processes of field ionization and field evaporation is also included. Descriptions of the components and the types of three-dimensional atom probe that have been developed are presented in chapter 4. The experimental factors that define the reliability of the three-dimensional ion-by-ion data are documented in chapter 5. The various methods to visualize and statistically analyze the three-dimensional data are described in chapter 6. A comprehensive list of the atom probe tomography applications is provided in a bibliography and the references at the end of each chapter.

This monograph is intended to be an experimentally oriented complement to the previously published textbook† on atom probe field ion microscopy. Therefore, some of the details of the theory of the physics behind the technique and the statistical analyses have been deferred to the comprehensive treatment of those topics in that book.

The text and figures for this monograph were provided in camera-ready form and therefore, all errors are the responsibility of the author and the idiosyncrasies of word processor used.

M. K. Miller, August 1999

† M. K. Miller, A. Cerezo, M. G. Hetherington and G. D. W. Smith, Atom Probe Field Ion Microscopy, Oxford University Press, Oxford, 1996.

Acknowledgments

The author would like to thank Drs. I. M. Anderson, S. S. Babu, J. A. Horton, E. A. Kenik, K. F. Russell, S. J. Sijbrandij and R. E. Stoller of Oak Ridge National Laboratory, Oak Ridge, TN for their valuable assistance in producing this monograph. The author would also like to express his sincere appreciation to the many people who have made significant contributions to this volume including Profs. D. Blavette and A. Menand and Drs. B. Deconihout, F. Danoix, F. Vurpillot, P. Pareige A. Bostel, E. Cadel, and P. Bas of the University of Rouen, Rouen, France, Dr. M. G. Burke of Bethel Bettis Inc., West Mifflin, PA, Dr. P. P. Camus of Noran, Middleton, WI, Dr. K. Hono of the National Research Institute for Metals, Tsukuba, Japan, Prof. T. F. Kelly, University of Wisconsin, Madison, WI, Dr. D. J. Larson, Seagate Technology, Bloomington, MN, Prof. M. Mousa, Mu'tah University, Al Kerak, Jordan, Prof. J. A. Panitz of University of New Mexico, Albuquerque, NM, and Prof. G. D. W. Smith, Drs. A. Cerezo and P. J. Warren of the University of Oxford, Oxford, England, Dr. R. C. Thomson of the University of Loughborough, Loughborough, England, and Dr. A. R. Waugh, Applied Vision, Coalville, England. The affiliations used in the figure captions are to the organization at which the research was originally performed. Many of the people have subsequently moved to other organizations and their current affiliations are as listed above. Several figures have been reprinted from various journals with the permissions of the publishers: Elsevier Science, Les Editions de Physique, IOP Publishing Ltd., The Royal Microscopical Society, and San Francisco Press, Inc., 660 Spruce Street, Berkeley, CA 94707.

The author would like to thank Amelia McNamara and her staff at Kluwer Academic/Plenum Publishers for their assistance in producing this volume.

This research was sponsored by the Division of Materials Science, U. S. Department of Energy, under contract DE-AXC05-96OR22464 with Lockheed Martin Research Corp. and through the ShaRE Program under contract DE-AC05-76OR00033 with Oak Ridge Associated Universities.

Contents

ATOM PROBE TOMOGRAPHY

Analysis at the Atomic Level

Chapter 1

Overview and Historical Evolution

One of the dreams in the characterization of materials is to be able to identify and determine the positions of all the atoms in a material with atomic precision. The technique of atom probe tomography permits this dream to be realized.

In this monograph, the technique is referred to as atom probe tomography (APT), since one of the definitions of tomography is a method of producing three-dimensional images of the internal structures of a solid object. The generic name of the instrument is a three-dimensional atom probe (3DAP). However, this name is slightly misleading, as all atom probes have three-dimensional characteristics.

In this chapter, the history of the development of the three-dimensional atom probe is presented together with a general introduction of the technique of atom probe tomography.

1.1 General Introduction

One of the major milestones in science and microscopy was the development of the field ion microscope. Ever since the concept of the atom was introduced in circa 450 BC by the Greek philosophers Democritus and Leucippus, scientists have attempted to image and identify individual atoms. At the present time, there are three types of microscopes that can achieve atomic resolution on a routine basis. They are the atom probe field ion microscope (APFIM), the high resolution electron microscope (HREM), and the scanning tunneling microscope (STM). Each of these instruments achieves atomic resolution by different methods and requires specimens with different geometries; the ionization and projection of image gas atoms by sharp needles for the field ion microscope, the interaction of an electron beam with thin foils for the electron microscope, and the movement of a sharp probe across relatively flat surfaces for the scanning tunneling microscope. The field ion microscope was the first instrument that allowed the resolution of individual atoms, predating the others by approximately 25 years.

The modern atom probe field ion microscope is still unique in its capabilities since it is the only technique that is currently available to see, select and identify an individual atom. Somewhat surprisingly, this ability is not the main advantage or use of the atom probe in its role as a tool for microstructural characterization. The most common use is to select and determine the composition of small volumes. Since the atom probe performs this task by counting the number of atoms of each element in that volume, it provides a fundamental measurement of the local composition.

The atom probe field ion microscope is used to perform ultrahigh resolution microstructural characterization of metals, semiconductors and some ceramics. Some typical applications of the atom probe are the quantification of the composition, size, morphology and number density of ultrafine precipitates, determination of the partitioning of alloying elements and impurities between the phases present in the microstructure, estimation of the segregation of elements to grain boundaries and other interphase interfaces and the detection and quantification of the early stages of phase separation.

1.2 Evolution of the Three-dimensional Atom Probe

On October 11th, 1955, Kenwar Bahadur and Prof. Erwin W. Müller at the Pennsylvania State University achieved the milestone of imaging individual atoms with the field ion microscope. An example of a portion of a field ion micrograph of a tungsten specimen is shown in Fig. 1.1. Each spot in this field ion micrograph is the image of a single atom. Several years later in 1967, Müller, Panitz and McLane were able to determine the identity of an atom with the atom probe. Today's state-of-the-art three-dimensional atom probe is one of the most powerful tools for the microstructural characterization of materials. The major steps that were involved in the development of the three-dimensional atom probe and the technique of atom probe tomography are described in the following sections.

1.2.1 Field electron emission

The atom probe traces it origins back to the field electron emission microscope (FEEM), also referred to as a field emission microscope (FEM) or sometimes the field electron microscope. This microscope is based on the principle that electron emission from a solid may occur in the presence of a strong electric field. The theory of electron emission from a solid by quantum mechanical tunneling was first considered by Fowler and Nordheim [1] and also by Oppenheimer [2] as early as 1928. Oppenheimer predicted that this

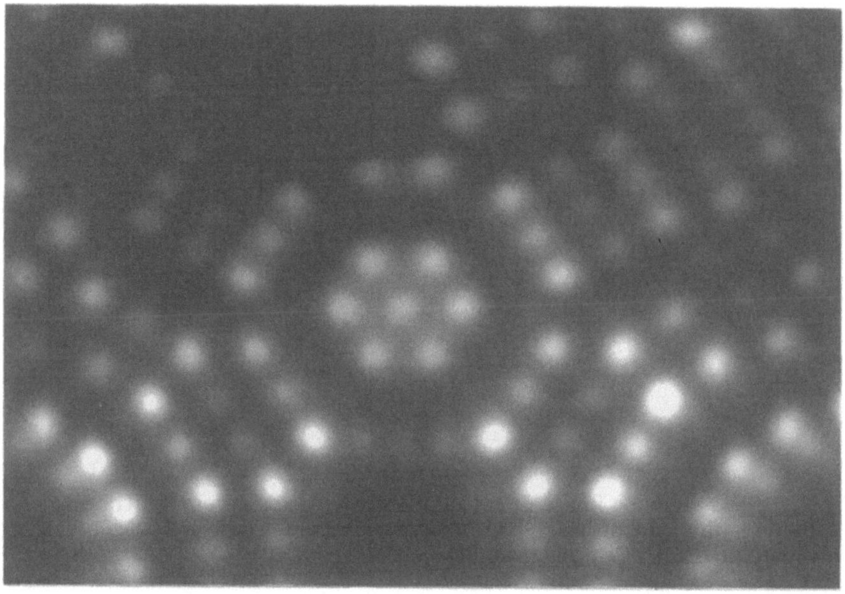

Fig. 1.1. Field ion micrograph of a 111-oriented tungsten specimen. In this field ion micrograph, each dot is the magnified image of an individual atom.

process requires an extremely high field strength of at least 10 V nm^{-1} to ionize a hydrogen atom from the ground state.

In 1935, Müller introduced a new type of microscope in which the specimen was in the form of a sharp needle and was placed in front of a phosphor screen in a vacuum system [3]. Since the field on the apex of the needle, F, is given by

$$F = \frac{V}{k\, r_t}, \quad 1.1$$

where V is the applied voltage, r_t is the end radius of the needle, and k is a numerical constant ($k \approx 2\text{-}5$), the use of a sharp needle produces a significant field enhancement factor and reduces the voltage required for ionization to experimentally attainable levels. With this arrangement, Müller was able to produce field strengths of the order of 1 V nm^{-1} with the application of only a few thousand negative volts to a needle with an end radius of ~1 μm. Under these conditions, an enlarged image of the specimen surface is projected on the phosphor screen, as shown in Fig. 1.2. The typical magnification of these field emission images is of the order of 10^4. The spatial resolution of the field

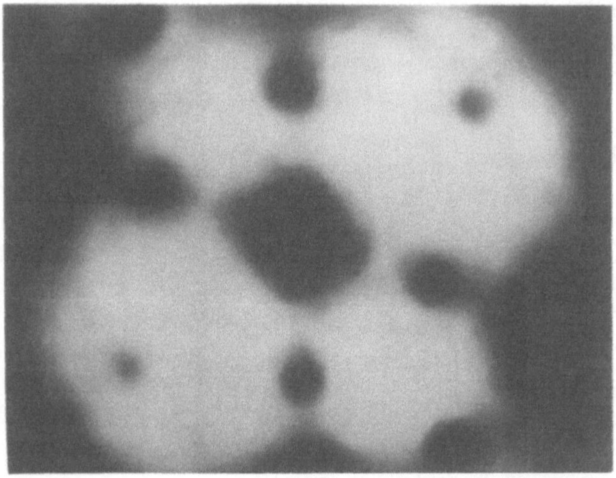

Fig. 1.2. Field emission image of a 110-oriented tungsten specimen. The spatial resolution in this type of image is of the order of 2 nm. Courtesy M. Mousa, University of Aston in Birmingham [4].

emission images was found to be of the order of 2 nm and although the performance was superior to other types of microscopes at that time, atomic resolution was not achieved in the field electron emission microscope.

Field emission sources are now routinely used as the high brightness source in the field emission guns of transmission and scanning electron microscopes. In addition, large area two-dimensional arrays of field emitters are being developed for high brightness, high resolution flat panel displays for the next generation of television monitors, etc.

1.2.2 Field ion microscope

Müller continued the quest for atomic resolution and in 1941 found that by reversing the bias on the emitter, the desorption of positively-charged ions could be obtained [5]. However, the quantity of these positively-charged ions was limited so a trace of hydrogen was admitted into the system to increase the number of ions emitted. In 1951, Müller published the first description of the field ion microscope [6]. In this instrument, the specimen was at room temperature but Müller stated that it would be an easy matter to cool the entire instrument by immersion into liquid air. This reduction in temperature was suggested to reduce the de Broglie wavelength even further than was provided with the use of hydrogen. Despite some earlier attempts with cooling the entire microscope to cryogenic temperatures, it was not until 11th October 1955 when Bahadur and Müller achieved atomic resolution field ion images of a tungsten specimen cooled with liquid nitrogen and with helium as the image gas. After this historic experiment was concluded, Müller remarked "Atoms, ja, atoms! "

A photograph of Prof. Müller with an early glass field ion microscope is shown in Fig. 1.3. In the early glass field ion microscopes, a phosphor was deposited on the inside of a large diameter glass port. Each specimen was mounted on a tungsten heating loop or filament and then this assembly was glass blown into the vacuum system. The heating loop was a legacy from the field emission microscope where the specimen was usually heated to high temperatures to remove contamination from the surface. At the time of these experiments, this heating process was the only method available to produce a clean surfacc, which was a necessary precursor to achieve atomic resolution. It is now known that this process also produces a thermally facetted end form and introduces some irregularities or disorder in the atomic arrangement in the surface atoms. Since this thermal conditioning process was applied to the specimens during the first few field ion microscopy experiments in 1951, atomic resolution images could not be obtained. The key to the eventual success was the use of higher field strengths that lead to the discovery of the process of field evaporation. This discovery was achieved through Bahadur's ability to produce extremely sharp needles and the switch to helium as the imaging gas. Since the specimens were extremely sharp, higher fields could be obtained at the apex of the needle. It is now known that to form a helium field ion image necessitates the application of a significantly higher field (~1.9 times) to the specimen compared to that required for hydrogen. Although Müller mentioned the use of helium in his 1951 paper, the first actual attempts to use helium (and other gases) were performed by Dreschler and Pankow in Berlin and the results were published in 1954 [7]. In a review paper in 1953, Müller describes a process of "tearing off" or removal of the surface tungsten atoms from the specimen at elevated temperature [8]. This is probably the first reference to the process now referred to as field evaporation. However, its importance was not fully realized at the time. In order to initiate field evaporation in tungsten, the field on the specimen has to be increased only slightly over that required to form a field ion image in helium. In contrast, almost double the field would be required if hydrogen is used as the imaging gas. In addition, when the voltage is increased in hydrogen, the image becomes blurred and extremely dim. Field evaporation was able to remove the surface layers of the specimen and thereby remove the artifacts introduced by the thermal heating process. Once these damaged layers were removed, atomic resolution field ion images were obtained by Bahadur and Müller. Field evaporation also provided an alternative method to produce an atomically clean surface and therefore eliminated the need to heat the specimen to elevated temperatures. The removal of surface atoms from the specimen may be carefully controlled with field evaporation, as shown in Fig. 1.4. In this

Fig. 1.3. Prof. E. W. Müller with an early glass field ion microscope. A ball model reconstruction of a field ion specimen may be seen in the background. Courtesy J. A. Panitz, University of New Mexico.

sequence of field ion micrographs, one atom is field evaporated from the atomic plane between each frame.

A schematic diagram of the key components of a field ion microscope is shown in Fig. 1.5. The needle-shaped specimen is attached to the base of a cryostat and pointed towards a phosphor screen. If the phosphor screen is positioned sufficiently close to the specimen, the entire field ion image can be observed and a specimen goniometer is not required. The specimen is also connected to a high voltage power supply to provide the required electrical field. Image gas can be admitted to the ultrahigh vacuum system at a controlled rate in a dynamically pumped system or to a specific pressure in a static vacuum. The intensity of these early field ion images on the phosphor screens was extremely low and the operator had to become completely dark adapted

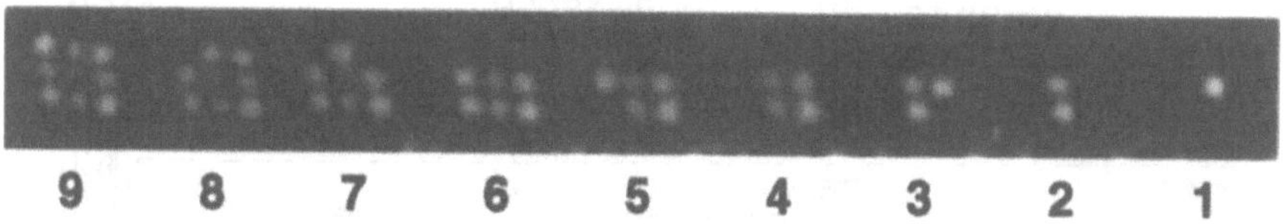

Fig. 1.4. Example of the precision that is possible with field evaporation. One atom is field evaporated from the surface of this nickel-zironium catalyst between each frame.

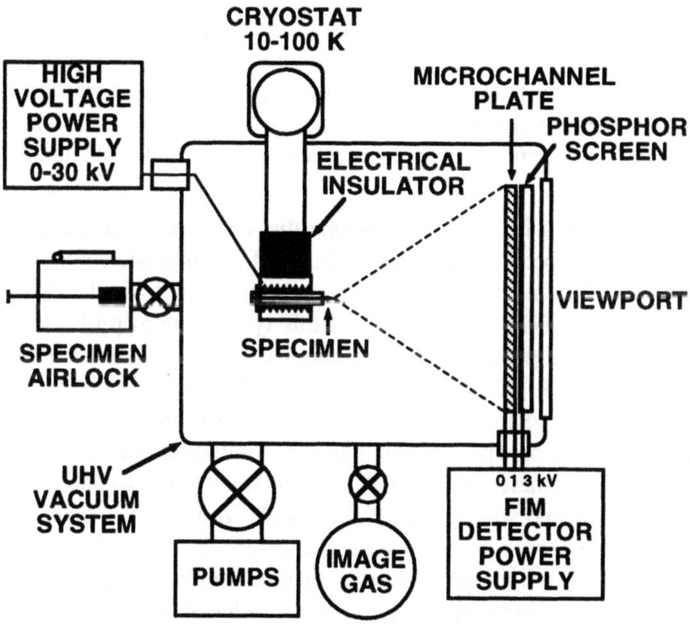

Fig. 1.5. Schematic diagram of the key components of a field ion microscope.

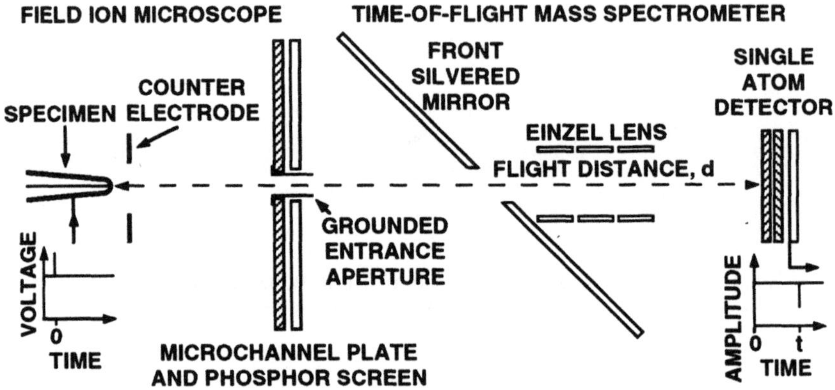

Fig. 1.6. Schematic diagram of a classical atom probe which features a linear time-of-flight mass spectrometer with a single atom detector.

before the start of an experiment. Sometimes tricks were employed to increase the brightness of the image. The introduction of microchannel plate image intensifiers (§4.2.5) in 1969 made a dramatic improvement in the ease of operation of a field ion microscope. The microchannel plate is positioned just in front of the phosphor screen, as shown in Fig. 1.5. The microchannel plate also permitted different image gases such as neon and argon to be used, which in turn significantly increased the range of materials that could be examined.

1.2.3 Atom probe

An atom probe is a field ion microscope that can analyze a specific atom or region with a mass spectrometer. The addition of a mass spectrometer to a field ion microscope enabled individual atoms to be identified for the first time. A schematic diagram of the key features of this type of atom probe is shown in Fig. 1.6. The original prototype atom probe was developed in 1967 by Müller, Panitz and McLane [9,10] and is shown in Fig. 1.7. The field ion microscope is at the top of this instrument and is mounted on a sliding glass seal so that the entire field ion microscope section can be rotated about the apex of the specimen. The goniometer enables any atom or small region visible in the field ion image to be aligned with a small circular hole in the center of the phosphor screen. This hole serves as the entrance aperture to the time-of-flight mass spectrometer. Although the physical size of this circular aperture is typically 1 to 2 mm, its effective size on the specimen is reduced by the magnification of the field ion microscope (i.e., 1 to 5 nm in diameter). When the specimen is field evaporated, atoms are removed from the entire surface of the specimen but only those that enter the mass spectrometer are analyzed, typically 0.1% of

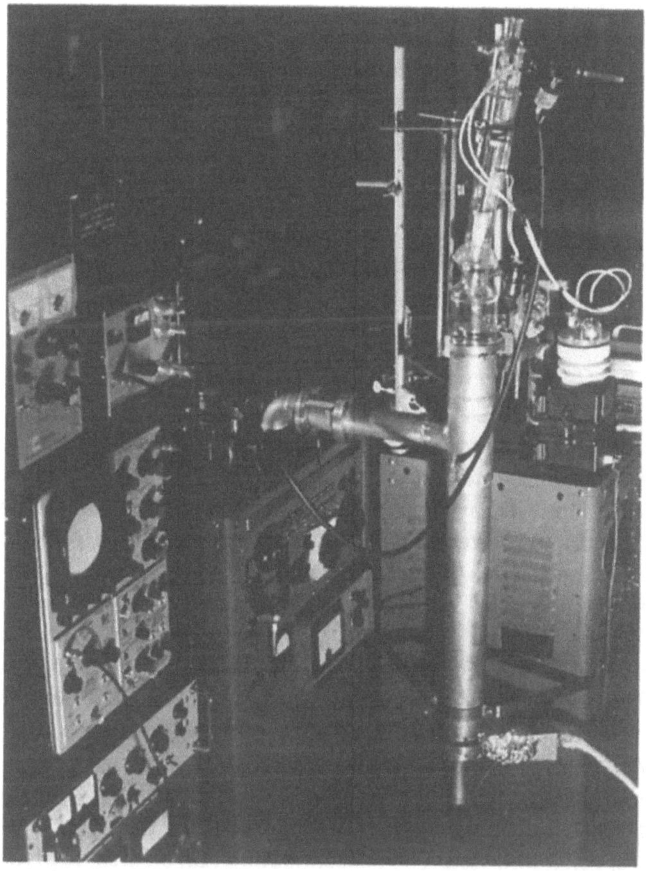

Fig. 1.7. The original atom probe developed by Müller, Panitz and McLane in 1967. The field ion microscope section is at the top of the instrument and the single atom detector at the end of the mass spectrometer is at the bottom. Courtesy J. A. Panitz, Pennsylvania State University, [10].

the total atoms evaporated. Therefore, this type of instrument is rather inefficient in the collection of the atoms field evaporated from the specimen. A single atom detector is positioned at the end of the time-of-flight mass spectrometer. The ions are removed from the specimen at a well defined time by superimposing a high voltage pulse on the standing voltage. In the original instrument, this pulse also started the trace on an oscilloscope. When an ion impacted the single atom detector, a deflection was observed on the oscilloscope trace. An example of a typical oscilloscope trace from this original instrument is shown in Fig. 1.8. The flight time of the ion, t, could be measured from this trace. The mass-to-charge ratio of the ion is determined, to a good first approximation, by equating the potential energy of the ion just prior to field evaporation to its kinetic energy just after field evaporation, i.e.,

$$n\,eV = \tfrac{1}{2} m\,v^2, \qquad 1.2$$

where n is the number of electrons removed in the field ionization process, e is the elementary charge, V is the total voltage on the specimen, m is the mass of the ion, and v is the velocity of the ion. Therefore by rearranging, the mass–to-charge ratio of the ion is given by

$$\frac{m}{n} = \text{constant}\; V \frac{t^2}{d^2}, \qquad 1.3$$

where d is the flight distance from the specimen to the single atom detector. The constant equals 0.1929796, if the voltage is expressed in kilovolts, the time in nanoseconds, the distance in millimeters, and the mass in atomic mass units. It is not possible to determine the mass independently from the charge. However, only one or two different charge states are normally observed for each element and this small number does not normally prevent the assignment of the ion to a particular element. In later instruments, the oscilloscope was replaced by a computer-controlled digital timing system (§4.3.4). By continuing this process over many field evaporation pulses, the number of ions collected at each mass-to-charge ratio can be accumulated into a mass spectrum. After the peaks in the mass spectrum are identified and assigned to elements (§5.4), the composition of the region analyzed can be calculated.

The composition of the volume is determined from the relative number of ions of each species collected. The concentration of an element is given by

$$c_i = \frac{n_i}{n_t}, \qquad 1.4$$

where n_i is the number of ions of element i, and n_t is the total number of ions collected. It should be noted that the atom probe provides a fundamental estimate of the composition and no correction factors are required. If the field evaporation process is continued for a large number of ions, atoms are collected from successive atomic layers and a volume of material is analyzed. This volume is usually referred to as the cylinder of analysis although its shape is more accurately described as a truncated cone due to the small increase in its diameter over the length of the experiment. The increase in diameter is due to the blunting of the needle-shaped specimen due to the removal of material during the experiment, which results in a progressive increase in the radius of

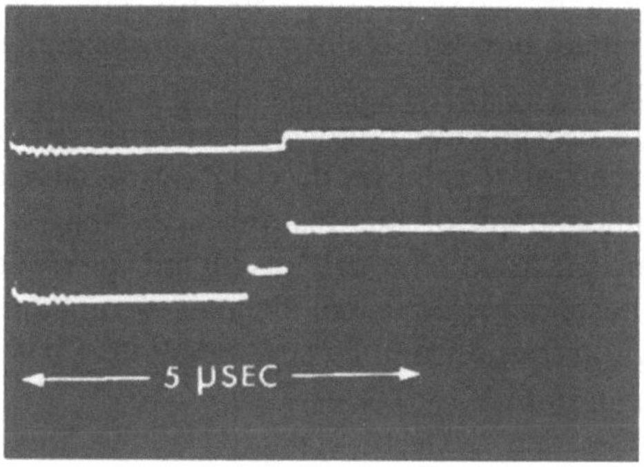

Fig. 1.8. A typical oscilloscope trace from an early atom probe. A vertical displacement indicates the arrival of an ion at the single atom detector. Courtesy J. A. Panitz, Pennsylvania State University.

the specimen and the consequent decrease in magnification. The variation in the concentrations of the solutes present can be examined by dividing this cylinder into small blocks of atoms and determining the compositions of each block. This procedure permits concentration variations as a function of distance to be determined on the atomic scale. Additional information may be obtained from these ion-by-ion data and some of the techniques are described in Chapter 6.

In contemporary instruments, either the relative position of the imaging screen containing the aperture can be moved along the ion-optical axis of the instrument or an iris is incorporated into the design so that the effective diameter of the aperture can be varied. Typical effective aperture sizes ranged from single atom dimensions to approximately 5 nm in diameter. It was also possible to maintain the same effective size of the aperture during an experiment by moving the aperture further from the specimen as it blunted due to field evaporation. This constant size enabled statistical methods, based on the analysis of time series, to be used in the analysis of the data. In addition, the mass resolution of this instrument was significantly improved by the incorporation of an energy-compensating lens in the mass spectrometer (§4.3.1).

1.2.4 Imaging atom probe

In order to increase the number of atoms analyzed from the specimen several new types of atom probe were developed. All these variants involved the elimination of the microchannel plate and screen assembly that contains the probe aperture to permit the single atom detector to be positioned closer to the specimen. Therefore, the area of analysis is defined by the active area of the single atom detector at the end of the mass spectrometer. Field ion microscopy could still be performed with this configuration by operating the single atom detector at low gain.

The first of these instruments, originally named the 10 cm atom probe and now known as the imaging atom probe (IAP), was developed by Panitz in 1973 [11,12]. A photograph of this instrument is shown in Fig. 1.9. The single atom detector consisted of a pair of curved microchannel plates and a matching curved phosphor screen where the radii of curvatures were chosen to ensure that the flight distance from the specimen to the detector was the same for all ions, as shown in Fig. 1.10.

In the main time-gating mode of operation of this instrument, the single atom detector was only turned on for a few nanoseconds at the precise time when a particular element would strike the detector. Each ion striking the detector would produce a visible dot on the detector so that the distribution of dots generated a two-dimensional map of that element over the specimen surface. The distribution of the other elements present could be obtained by adjusting the time at which the detector was turned on. The instrument could also be used to obtain a mass spectrum of the surface of the specimen on a single field evaporation pulse. Panitz also pioneered the use of video cameras on the imaging atom probe to routinely record the images [13].

The imaging atom probe did not find significant use in metallurgical investigations since the dot maps could not be quantified in terms of local composition. In addition, it is not possible to correlate the maps of different elements since each map is taken from a different shell of material field evaporated from the specimen and is therefore at a slightly different position along the specimen axis. However, the imaging atom probe provided some valuable insight about the trajectory aberrations of field emitted ions (§3.2.9).

Panitz used the imaging atom probe to investigate biological molecules deposited onto the apex region of the field ion specimen [14]. In this technique, which was called field ion tomography (FIT), a thin layer of benzene (or water) ice was condensed from benzene (or water) vapor on top of the cryogenically-cooled unstained biological molecules. The layers of benzene ice were then field desorbed and the images of the benzene ions on the detector were combined and reconstructed in a computer. An example of a three-dimensional reconstruction of a ferritin molecule is shown in Fig. 1.11. Since no benzene ions were desorbed from the ferritin molecule, the spherical shape of the molecule is apparent in the reconstruction.

1.2.5 Three-dimensional atom probe

The original concept of the three-dimensional atom probe was developed by the author between 1983 and 1986 [15] and was to a large extent inspired by a comment made by Panitz in his patent on the imaging atom probe about an

Fig. 1.9. Photograph of the 10 cm atom probe developed by Panitz. Courtesy J. A. Panitz, Sandia National Laboratory.

instrument that could make "atom-by-atom analysis at several different locations simultaneously" [12]. The first prototype of a three-dimensional atom probe was implemented in early 1986 by the author. This instrument was based on an imaging atom probe configuration with the addition of two secondary detectors that were external to the vacuum system, as shown in Fig. 1.12. Both secondary detectors were focussed on the light output from the imaging atom probe detector with the use of a beam splitter and optical lenses. The first detector was a CCD camera that was used to determine the x and y coordinates of the ions striking the detector. One of the first examples of the output of the CCD camera for an ion striking the detector from this prototype instrument is shown in Fig. 1.13. The second detector was a 32 x 32 array of photodiodes to determine the flight times with the use of 1024 channels of time-to-digital converters. Each photodiode was designed to map to a 4 x 4 subset of pixels on the CCD camera. Although the feasibility of this secondary detector was demonstrated with individual photodiodes, the full photodiode array was never implemented. However, initial experiments with this instrument indicated that the signal level on the CCD camera was marginal and resulted in having to operate the primary detector at extremely high gain. In addition, this prototype suffered from excessive background noise on the camera detector. This approach was to reemerge a few years later as the optical atom probe (OAP) [16,17] when the sensitivity and resolution of CCD cameras

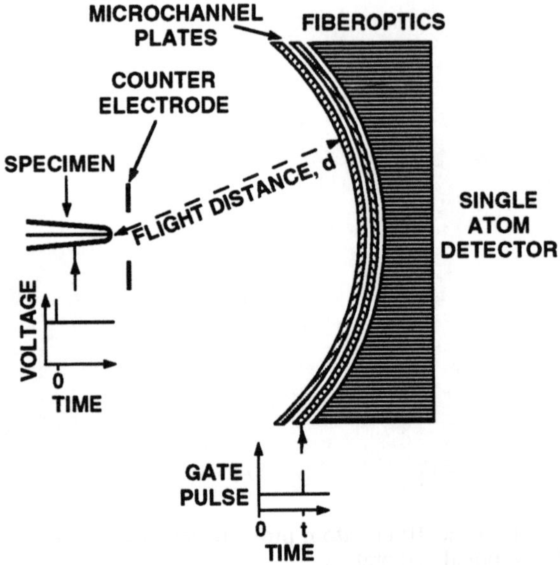

Fig. 1.10. Schematic diagram of the imaging atom probe. The microchannel plates and screen of the single atom detector are curved so that the flight distance is the same to all points on the detector. The microchannel plates may be time-gated to record the arrival of a single species.

had improved (§4.3.3) and Kellogg had introduced a variant of the imaging atom probe with a time-gated image intensifier [18].

During the same period, Schiller et al. implemented a computerized imaging system for field ion microscopy and time-gated time-of-flight imaging [19]. This single atom sensitive detector was based on a pair of microchannel plates and a resistive anode. The x and y coordinates of each image gas ion striking the detector were determined from the relative electrical charges measured at the four corners of the resistive anode. Although no attempts were made to use this instrument as an atom probe, this instrument incorporated all the essential features of a three-dimensional atom probe.

The first fully operational instrument, known as the position-sensitive atom probe (PoSAP), was developed by Cerezo et al. at the University of Oxford in 1988 [20]. This instrument featured a single atom detector based on a pair of microchannel plates and a wedge-and-strip or backgammon anode [21]. The x and y coordinates of the ion's impact on the detector are determined from the relative charge measured on the three anodes in the wedge-and-strip detector. Additional details of the position-sensitive atom probe are given in §4.4.1.

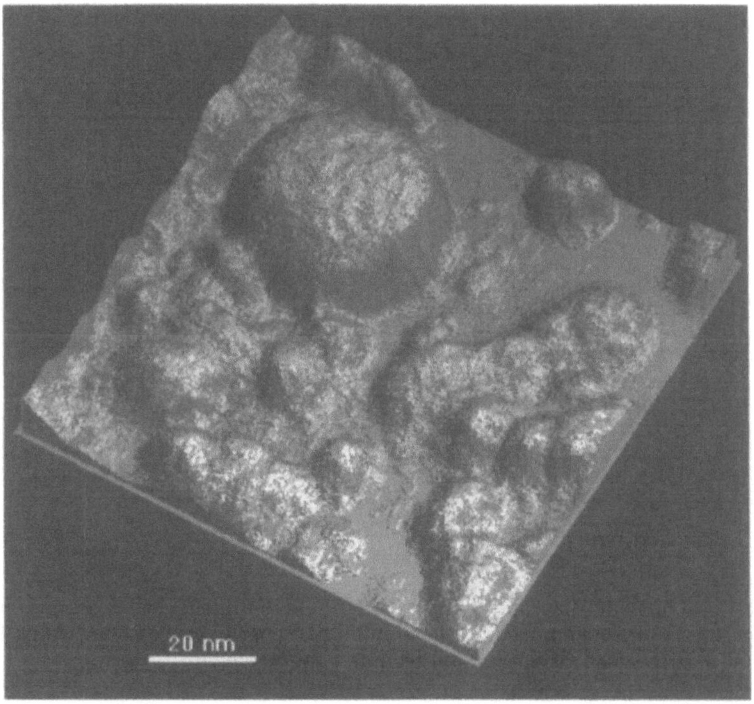

Fig. 1.11. Three-dimensional reconstruction of a ferritin molecule obtained by field ion tomography [14]. Courtesy J. A. Panitz, Sandia National Laboratory.

A variety of position-sensitive detectors have subsequently been developed including detectors based on CCD or video cameras, and multianode arrays. These different designs have resulted in a variety of names and acronyms (optical atom probe (OAP) [17], tomographic atom probe (TAP) [22], optical position-sensitive atom probe (OPoSAP) [23] and optical tomographic atom probe (OTAP) [24]). A detailed description of differences in these variants is given in Chapter 4. Some of these instruments and their single atom detectors, namely the position-sensitive atom probe, the tomographic atom probe, and the optical position-sensitive atom probe are commercially available.

In all these instruments, the spatial coordinates and the mass-to-charge ratios of the atoms within the analyzed volume are determined. A photograph of a state-of-the-art energy-compensated three-dimensional atom probe is shown in Fig. 1.14. A schematic diagram of a generic three-dimensional atom probe is shown in Fig. 1.15. The microscope consists of the needle-shaped

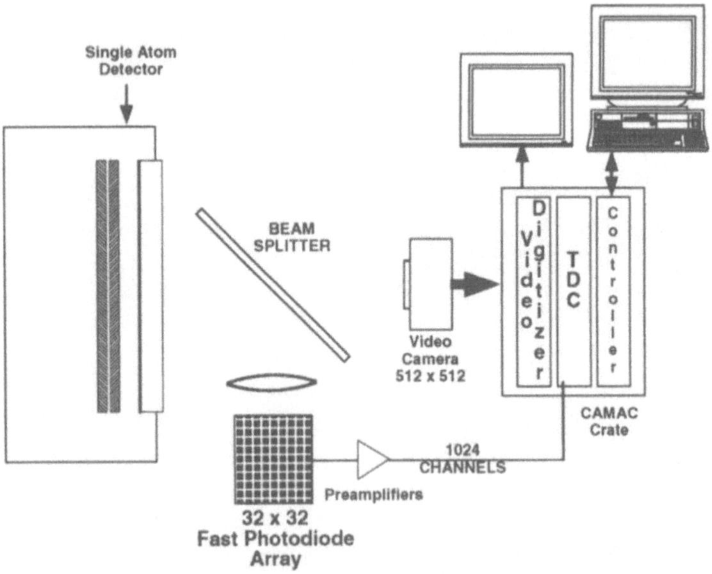

Fig. 1.12. Schematic diagram of the prototype three-dimensional atom probe. This prototype includes a CCD camera and a photodiode array to determine the position of the impact of the ion on the single atom detector.

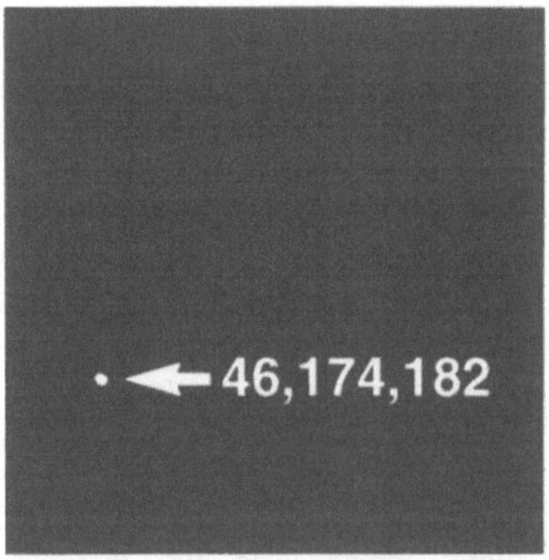

Fig. 1.13. Example of the output on the CCD camera for a single ion striking the detector obtained on the prototype optical atom probe in 1986. The x and y coordinates were determined to be 46 and 172 pixels, respectively and the maximum intensity was 182.

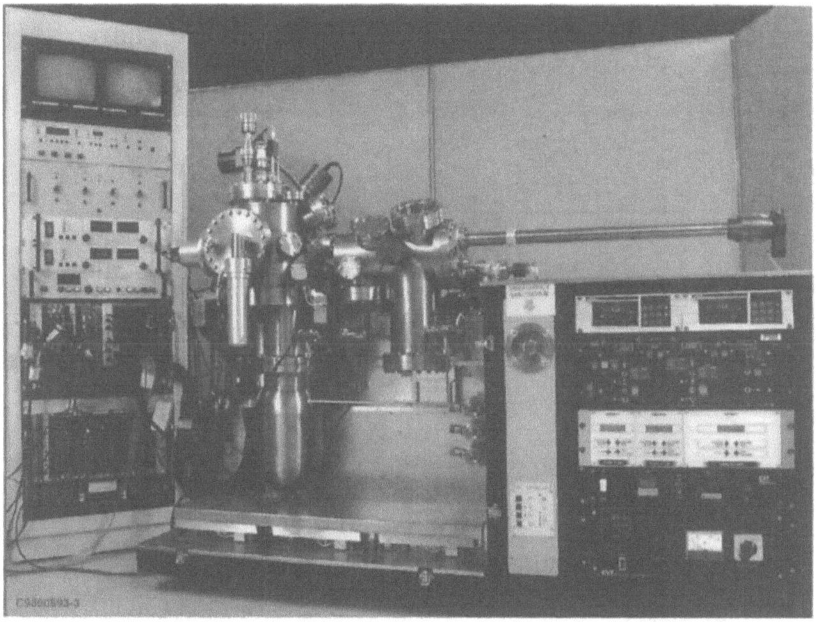

Fig. 1.14. A photograph of a commercial energy-compensated three-dimensional atom probe.

field ion specimen mounted on a cryogenically cooled goniometer. A circular counter electrode is positioned close to the apex of the specimen. A single atom position-sensitive detector is positioned at the end of the time-of-flight mass spectrometer. The typical distance from the detector to the specimen is between 250 and 650 mm. Details of all the components that are used in this instrument are discussed in §4.4. In this instrument, the active area of this detector defines the area analyzed on the specimen. It should be noted that a three-dimensional atom probe may be operated as a classic atom probe by the incorporation of an aperture in the mass spectrometer and ignoring the spatial information from the single atom detector. Alternatively, the typical cylinder of analysis obtained in a classic atom probe may be reconstructed from the three-dimensional data. Although field ion images of the area to be analyzed may be formed directly on the position-sensitive detector, a separate microchannel plate and phosphor screen assembly is generally used so that the field ion image of the entire specimen, which is significantly larger than the analysis area, may be examined.

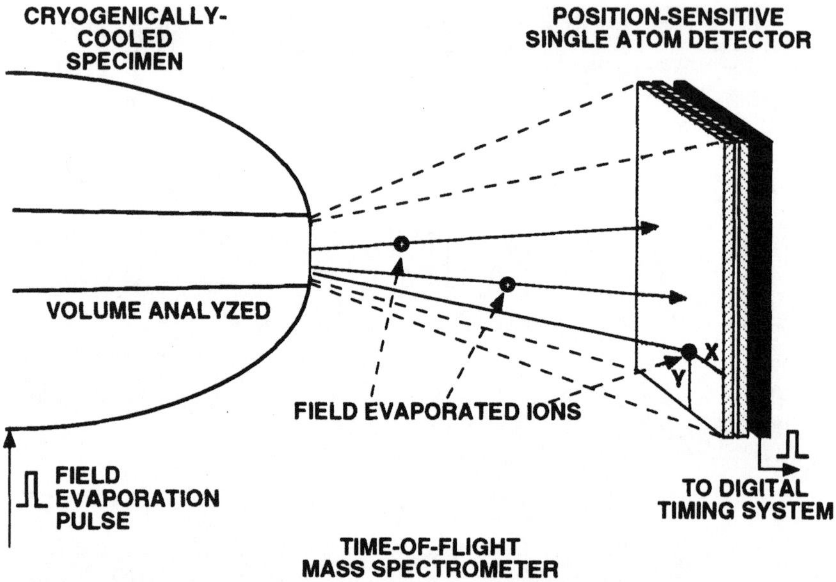

Fig. 1.15. Schematic diagram of a three-dimensional atom probe. The active area of the single atom detector defines the area of analysis.

When an ion is field evaporated in a three-dimensional atom probe, its mass-to-charge ratio is determined from its flight time from the specimen to the position-sensitive detector, as in previous atom probe designs (i.e., eqn 1.3). The x and y coordinates of its impact position on the two-dimensional position-sensitive detector, X_a and Y_a, are used to estimate two of the atom coordinates in the specimen, X and$Y$, with the relationships

$$X = \frac{X_{\rm d}}{\eta} \quad and \quad Y = \frac{Y_{\rm d}}{\eta}, \qquad 1.5$$

where the magnification, η, is given to a first approximation by

$$\eta = \frac{d}{\xi\, r_{\rm t}} \qquad 1.6$$

where d is the specimen to detector distance, ξ is a projection parameter known as the image compression factor (§3.3.3), and $r_{\rm t}$ is the radius of curvature of the specimen. Additional details of the reconstruction of the atom positions are given in §5.5. The z atom coordinate is determined from the position in the evaporation sequence in the same manner as used to determine the depth scale

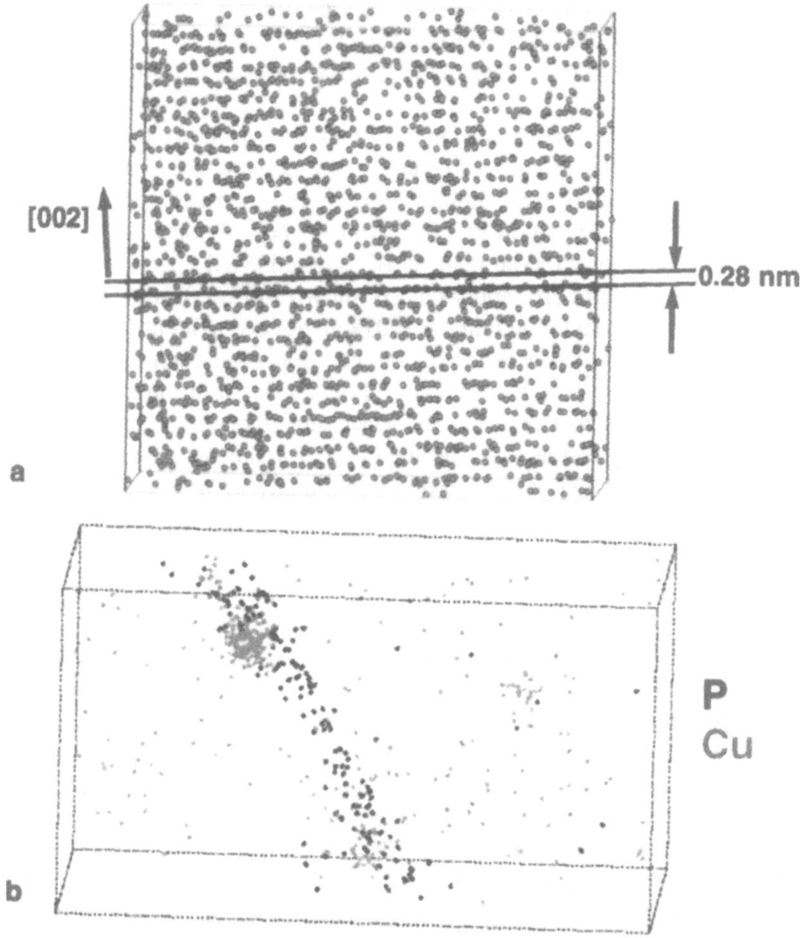

Fig. 1.16. Atom map of a) the distribution of aluminum atoms in a B2-ordered NiAl intermetallic specimen and b) a dislocation in a neutron-irradiated Cr-Mo-V pressure vessel steel that exhibits both phosphorus decoration and a copper-enriched precitpitate. Courtesy S. Duval and P. Pareige, University of Rouen.

in the composition profile in a probe aperture type of atom probe (§5.5.2). A typical sample volume in a three-dimensional atom probe is an area of ~10 x ~10 nm to ~20 x 20 nm by a depth of ~100 to ~250 nm and contains approximately 10^5 to 10^6 atoms. The time taken to accumulate this number of atoms is typically of the order of 20 h. Due to this extended time, the components and the software controlling a three-dimensional atom probe are generally configured to permit unattended operation after the initial setting up period.

These data are then reconstructed in a computer to visualize and characterize the arrangement of the atoms, to determine composition variations, and to estimate other parameters. The simplest type of representation is the atom or dot map in which the position of each atom in the volume is as a dot or a sphere. An example of an atom map of B2-ordered NiAl specimen in which only the aluminum atoms are represented is shown in Fig. 1.16a. The volume has been oriented so that the aluminum (001) planes are evident. A dislocation that exhibits both phosphorus decoration and a copper-enriched precipitate in a Cr-Mo-V pressure vessel steel is shown in Fig. 1.16b. Additional examples of the types of analyses that may be performed are given in Plates 1-X. The techniques to display and analyze the three-dimensional data are described in detail in Chapter 6. A list of atom probe tomography applications is given in the Bibliography.

References

1. R. H. Fowler and L. W. Nordheim, *Proc. Trans. Roy. Soc. Lond. A*, **119** (1928) 173.
2. J. R. Oppenheimer, *Phys. Rev.*, **31** (1928) 67.
3. E. W. Müller, *Zh. Tekh. Fiz*, **17** (1936) 412.
4. R. Latham and M. Mousa, *J. Phys. D Appl. Phys.*, **19** (1986) 699.
5. E. W. Müller, *Naturwiss.*, **29** (1941) 533.
6. E. W. Müller, *Z. Phys.*, **31** (1951) 136.
7. M. Dreschler and G. Pankow, Proc. 3rd Intl. Conf. Electron Microscopy, London, UK, July 15-21, 1954, V.E. Cosslett, ed., The Royal Microscopical Society, London, UK, (1956) 405.
8. E. W. Müller, *Ergeb. Exakten Naturwiss.*, **27** (1953) 290.
9. E. W. Müller, J. A. Panitz and S. B. McLane, *Rev. Sci. Instrum.*, **39** (1968) 83.
10. J. A. Panitz, *Crit. Rev. Solid State*, **5** (1975) 153.
11. J. A. Panitz, *Rev. Sci. Instrum.*, **44** (1973) 1034.
12. J. A. Panitz, Field desorption spectrometer, US Patent No. 3,868,507 (1975).
13. J. A. Panitz, *J. Vac. Sci. Technol.*, **17** (1980) 757.
14. J. A. Panitz, Proc. 52nd Annual Meeting Electron Microscopy Society, July 31- Aug. 5, 1994, New Orleans, LA, G. W. Bailey and A. J. Garrett-Reed, eds., San Francisco Press, San Francisco, CA, (1984) 824.

15. M. K. Miller, presented at the 21st Microbeam Analysis Society meeting, Albuquerque, NM, 1986.
16. M. K. Miller, *Surf. Sci.,* **246** (1991) 428.
17. M. K. Miller, *Surf. Sci.,* **266** (1992) 494.
18. G. L. Kellogg, *Rev. Sci. Instrum.*, **58** (1987) 38.
19. T. Schiller, U. Weigmann, S. Jaenicke and J. H. Block, *J. de Phys.*, **47-C2**, (1986) 479.
20. A. Cerezo, T. J. Godfrey and G. D. W. Smith, *Rev. Sci. Instrum.*, **59** (1988) 862.
21. C. Martin, P. Jelinsky, M. Lampton, R. F. Malina and H. O. Auger, *Rev. Sci. Instrum.*, **52** (1981) 1067.
22. B. Deconihout, A. Bostel, A. Menand, J. M. Sarrau, M. Bouet, S. Chambreland and D. Blavette, *Appl. Surf. Sci.*, **67** (1993) 444.
23. A. Cerezo, T. J. Godfrey, J. M. Hyde, S. J. Sijbrandij, and G. D. W. Smith, *Appl. Surf. Sci.*, **76/77** (1994) 374.
24. B. Deconihout, L. Renaud, G. Da Costa, M. Bouet, A. Bostel and D. Blavette, *Ultramicroscopy*, **73** (1998) 253.

Plate Captions

Plate I. Atom map of a boron-doped (400 appm) B2-ordered Fe-40 at. % Al material showing the presence of a boron-enriched Cottrell atmosphere along an edge dislocation [1]. The dislocation line is perpendicular to the figure and the Burgers vector (a<001>) is perpendicular to Al-rich (001) planes. Courtesy D. Blavette, University of Rouen, France.

Plate II. Atom map of a $Ti_{48}Al_{48}Cr_2Nb_2$ alloy aged for 96 h at 1000°C [2]. Only oxygen (green) and chromium (blue) atoms are represented. The oxygen atoms are preferentially located in the α_2 phase. Chromium segregation is evident at both α_2 / γ and γ / γ interfaces. Courtesy A. Menand, University of Rouen, France.

Plate III. Atom map of Ω phase in an Al-1.9% Cu-0.3% Mg-0.2% Ag alloy aged for 10 h at 180°C [3]. The precipitate-matrix interface is enriched in magnesium and silver. A segregant-free 2.76-nm-high ledge is also evident. Courtesy K. Hono, National Research Institute for Materials, Tsukuba, Japan.

Plate IV. Atom map of Ω and θ' phases in an Al-1.9% Cu-0.3% Mg-0.2% Ag alloy aged for 10 h at 180°C in which the {111} atomic layers are resolved [3]. The inclined θ' platelet is on the (100) plane. This result demonstrates that precipitates on different habit planes can be reconstructed accurately in atom maps. Courtesy K. Hono, National Research Institute for Materials, Tsukuba, Japan.

Plate V. Grain boundary segregation in a neutron-irradiated weld from a Russian VVER 440 nuclear reactor. Enrichments of phosphorus, molybdenum, and manganese are evident at the boundary in this atom map. Three ultrafine copper-enriched precipitates near the boundary are also evident.

Plate VI. X-ray tracer representation of copper-enriched precipitates in an Fe-Cu model alloy. The spectrum of colors indicates the local maximum copper concentrations, with red corresponding to the maximum and blue the minimum concentration. From a collaboration with P. Pariege and D. Blavette, University of Rouen, France.

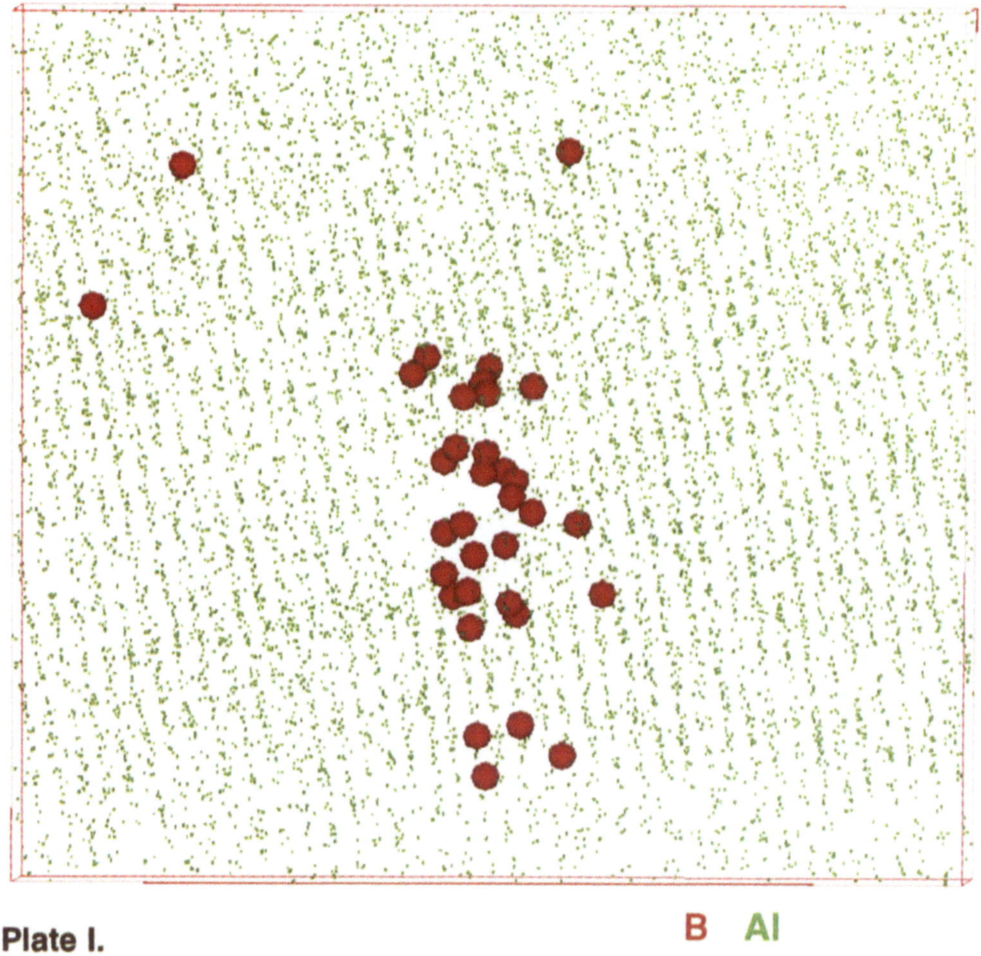

Plate I.

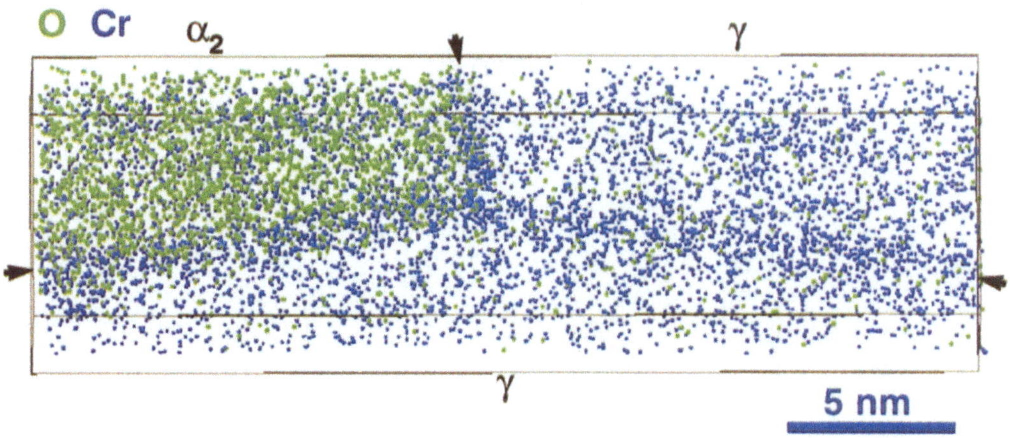

Plate II.

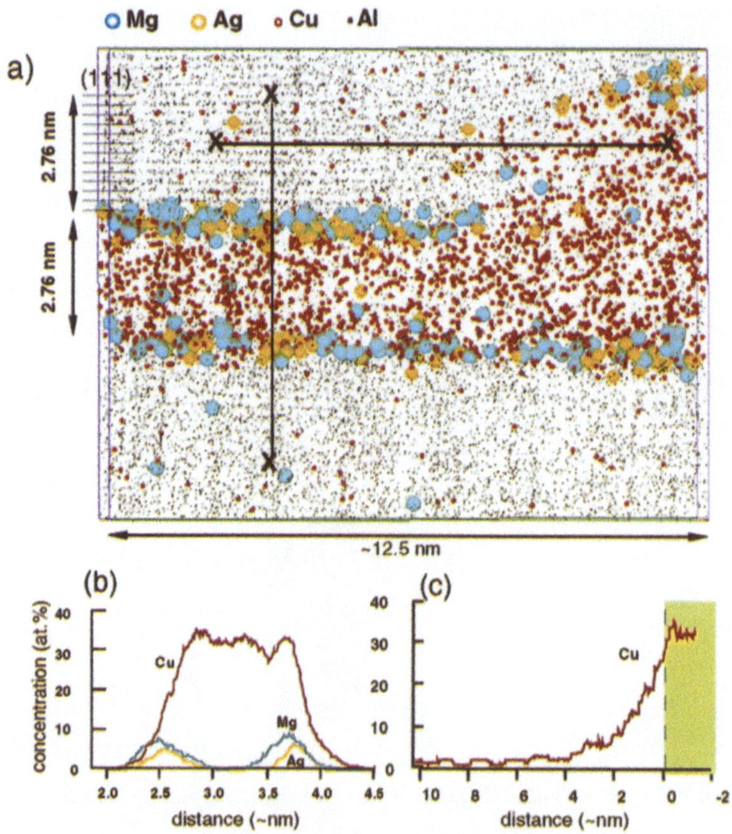

Plate III.

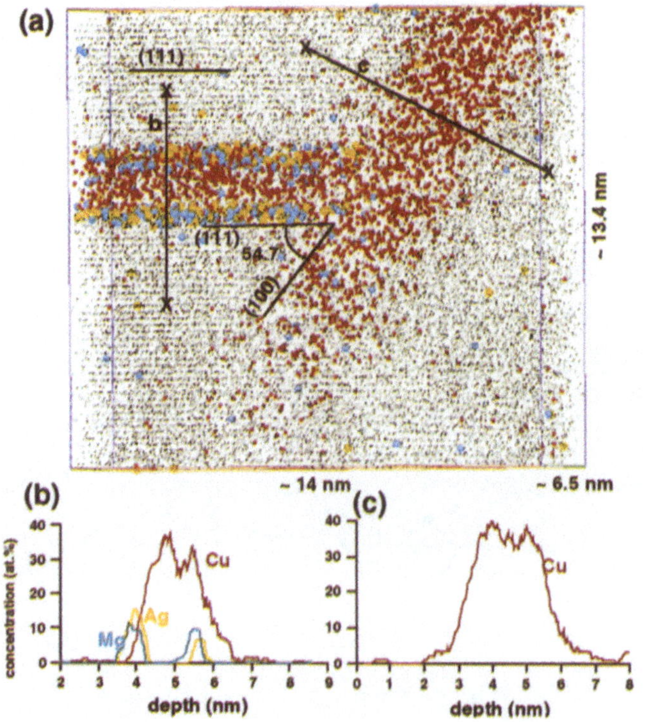

Plate IV.

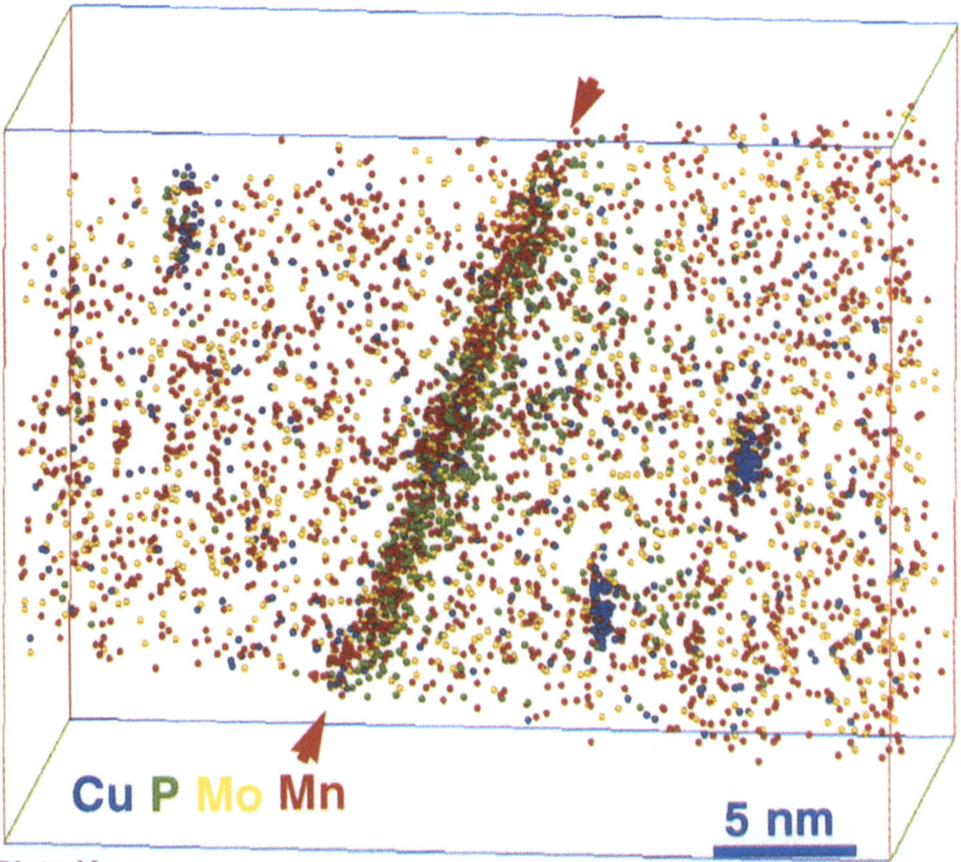

Plate V.

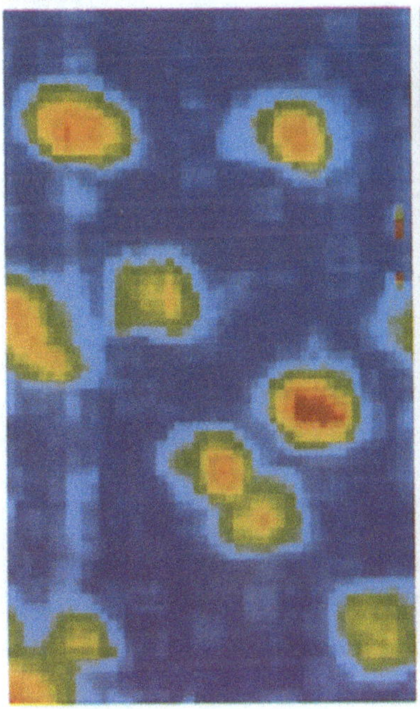

Plate VI.

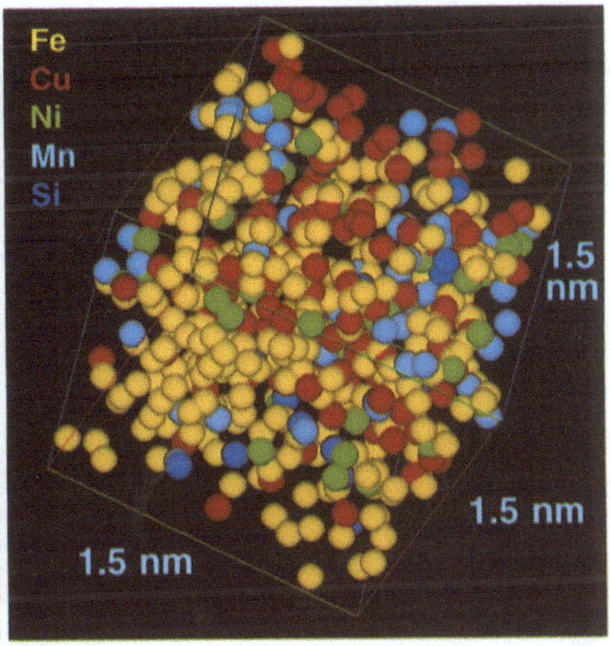

Plate VII.

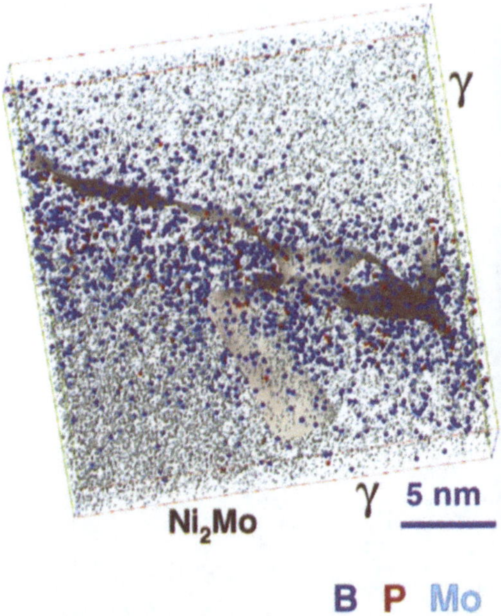

Plate VIII.

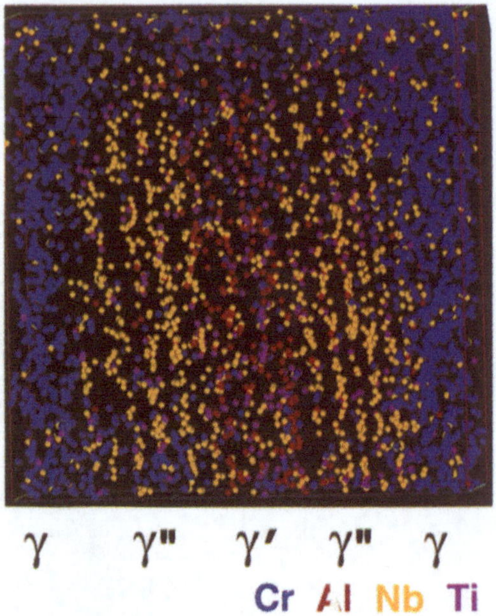

Plate IX.

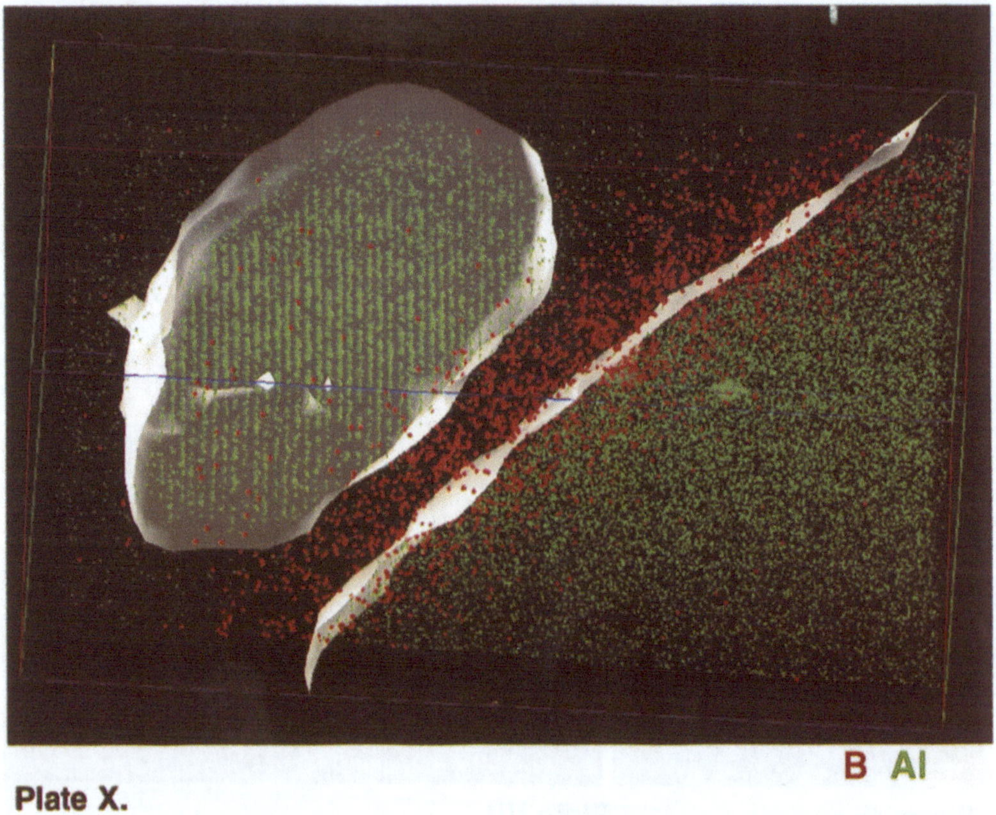

Plate X.

Plate VII. The solute distribution at the center of a 2-nm-diameter copper-enriched precipitate in a neutron-irradiated A55B pressure vessel steel. All atoms are shown. Enrichments of nickel, manganese, and silicon and a high level of iron are evident.

Plate VIII. Atom map of a boron- and phosphorus-enriched film at a grain boundary in HAYNES® 242™ nickel base superalloy. The enrichment is evident at both γ–γ and Ni_2Mo-γ segments of the boundary. The molybdenum isoconcentration surface (16% Mo) also indicates molybdenum enrichment at the γ-γ segment of the boundary.

Plate IX. Section of an atom map through a secondary precipitate in a nickel based superalloy Alloy 718. This precipitate exhibits two regions of the γ′ $L1_2$-ordered phase and a central region of the γ″ DO_{22}-ordered phase [4].

Plate X. Grain boundary segregation of boron (red atoms) in nickel base superalloy N18 [5]. The aluminum isoconcentration surfaces depict the interfaces between the γ′ precipitates and the γ matrix. The three-dimensional reconstruction has been oriented to show the aluminum-enriched (green atoms) (001) planes in the smaller $L1_2$-ordered γ′ precipitate. Courtesy E. Cadel and D. Blavette, University of Rouen, France.

References

1. D. Blavette, E. Cadel, A. Fraczkiewicz and A. Menand, *Science*, in press.
2. A. Menand, E. Cadel, C. Pareige and D. Blavette, *Ultramicroscopy*, **78** (1999) 63.
3. K. Hono, *Acta Mater.*, **47** (1999) 3127.
4. M. K. Miller, S. S. Babu and M. G. Burke, *Mater. Sci. Eng.*, **A270** (1999) 14.
5. D. Blavette, G. Geandier, E. Cadel, F. Danoix and A. Menand, *J. de Phys. IV*, **9** (1999) 113.

Chapter 2

The Art of Specimen Preparation

Atom probe field ion microscopy has been applied to a wide variety of materials ranging from pure metals and simple model alloys to complex multicomponent engineering alloys. Suitable materials include most metals, semiconductors and some ceramics. The main requirement is that the material exhibits some electrical conduction. However, thin insulating films may also be characterized. Field ion microscopy has also been applied to thin metallic films deposited on either needle-shaped or flat substrates. Other requirements are that the specimen is a solid at room temperature that does not have a high vapor pressure and has sufficient mechanical strength to be self supporting. Some of these unsuitable materials, such as gallium, are used to form ion beams in the related field of liquid metal ion sources.

In order to attain the field strengths of 20 to 40 Vnm^{-1} required to ionize both the atoms on the specimen surface and image gas atoms at reasonable applied voltages (typically 5 to 20 kV), a needle-shaped specimen with an end radius of 10 to 100 nm is required. The specimen must have a smooth or polished surface finish and be free from any protrusions, grooves or cracks, as shown in a rapidly-solidified Ni_3Al specimen [1] in Fig. 2.1. Since material is removed from the specimen during an analysis, a specimen that has a hemispherical end cap on a parallel-sided cylindrical base is ideal. However, this geometry is difficult, if not impossible, to achieve routinely in practice and a typical specimen has a hemispherical cap on a truncated cone with a taper angle of up to a half angle of 15°. This small taper angle causes the end radius of the specimen to increase during analysis as material is field evaporated. Therefore, the voltage on the specimen has to be increased to maintain the same field strength. Smaller taper angles are desirable, since the distance field evaporated along the specimen axis is greater for smaller taper

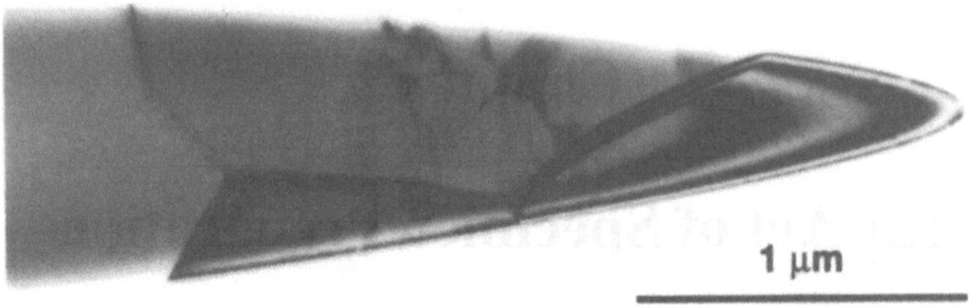

Fig. 2.1. Field ion specimen of boron-doped Ni_3Al containing grain and antiphase boundaries. The end radius of the needle is ~50 nm. The dark lines in the apex region are thickness fringes due to the circular cross section of the needle. Courtesy J. A. Horton, Oak Ridge National Laboratory [1].

angles for the same initial and final specimen radii. Therefore, more atoms are collected from the specimen. The typical length of the needle is ~5 mm and is mainly dictated by the ease of handling requirements. In most instruments, the field ion specimen is mounted on a support structure either by crimping in a copper or nickel tube or adhering to a wire pin with an electrically conducting paste or epoxy, as discussed in §4.2.1. If the length of the needle is less than ~1 mm, the full field strength may not be produced at the apex of the specimen due to shielding from the support structure.

The methods used to fabricate suitable field ion specimens with various techniques are outlined in the remainder of this Chapter.

2.1 Initial Preparation Methods

Many diverse methods have been developed to fabricate specimens suitable for field ion microscopy [2-4]. It should be noted that these methods may also be applied to fabricate needle-shaped specimens and probes for other techniques such as scanning tunneling microscopy.

The first stage of this specimen preparation process is to produce a square cross section blank that is typically 0.2 to 0.5 mm square by ~10 mm in length. This blank is generally cut out of the bulk material with a water-cooled diamond impregnated cutting wheel or a water-cooled wire saw. It is important that the initial cross section of the blank is uniform, since any deviation from a square or round cross section will normally produce an undesirable wedge-shaped needle. Centerless grinding may be used to produce round bars. Spark machining and related methods may also be used to produce round or square bars. It should be noted that the surface damage introduced by these initial preparation stages will normally be removed during subsequent electropolishing. Wire drawing and swaging may also be used if the deformation introduced does not interfere with the microstructure or microchemistry of the specimen.

Other more specialized techniques such as splat-quenching, injecting a molten stream of material through a small orifice into water or other liquid, in-rotating-water spinning, and pendant drop melt extraction have also been used in the examination of rapidly-solidified materials [1,5-11]. Some other forms of starting material such as whiskers or fibers may not require these initial stages. Special procedures are required for thin film materials and these are discussed in the next section.

2.2 Thin Films

The fabrication of field ion specimens from thin films and multilayers requires special techniques to ensure that the film is located in the small volume of material at the apex region of the specimen that can be analyzed. One method is to deposit the thin film on the end of a field evaporated needle. This in situ method has the advantage that the film may be deposited onto an atomically clean substrate under well-defined conditions of temperature and ultrahigh vacuum. This operation can be performed in the preparation chamber of the atom probe after the specimen has been field evaporated in the field ion microscope. The hemispherical apex region of the needle provides a suitable surface to examine film formation on different crystallographic facets. However, the highly curved surface may alter the deposition and growth characteristics of the film and so it may not be representative of films grown on flat surfaces. It should be noted that depositing films on needle substrates will increase the end radius and therefore there is a practical limit to the thickness of the film that can be deposited.

Lithography-based techniques have been developed to fabricate blanks from thin films deposited onto flat substrates [12-15]. These methods have the advantage that they may be applied to thicker films that cannot be fabricated by deposition onto needle-shaped substrates. The first step of this process is to fabricate the thin film on a substrate coated with a suitable removable underlayer. The choice of the material for the underlayer is critical since it controls the adhesion of the film to the substrate, may be used to prevent reactions between the film and the substrate, and facilitates the removal of the film from the substrate. In addition, a seed layer is often required to control the crystallographic orientation and crystal structure of the film. In practice, multiple layers may be necessary to fulfill these requirements. A photoresist is then applied to the exposed surface of the thin film. The photoresist is exposed through a mask delineating the desired specimen blanks [12]. A typical mask is reproduced in Fig. 2.2a. This mask features a thin parallel-sided region where the field ion needle will be formed and a thicker blade-shaped region (~200 by ~200 μm) that is used to handle the blank in the later stages of this process. The width of the thin parallel-sided region is chosen to match the thickness of the film. The unwanted regions of thin film are then removed by ion milling until the underlayer is exposed. Next, the underlayer is chemically removed to separate the specimen blanks from the substrate. Thousands of specimen blanks may be produced from one thin film sample in a single run. The specimen blank is then attached with an electrically conducting paste to a metallic support wire, such as tungsten, to facilitate handing in the subsequent operations, as shown in Fig. 2.2b. Due to the small size of the blank, this step has to be performed with the aid of an optical microscope and special specimen handling equipment. The final step in the process is to micropolish the square parallel-sided region of the specimen blank to form a suitable field ion needle (§2.3.3). It should be noted that the specimen axis of the final needle is in the planar direction of the film and the analyzable region is in the central portion of the film. Other methods based on ion milling have been developed to examine the near surface regions (§2.4.2 and §2.4.3).

2.3 Electropolishing

By far the most common method used to produce the desired needle-shaped specimen from metallic blanks is electropolishing. For each material, the electropolishing conditions are defined by a number of experimental parameters including the composition, viscosity and temperature of the electrolyte and the voltage and current applied to the cell. A schematic diagram of a voltage/current density curve for a typical electropolishing solution is shown in Fig. 2.3. The curve is divided into three sections. In the low voltage regions AB, the more reactive points on the surface of the needle are preferentially attacked. This electrolytical etch region does not usually produce a good or shiny surface finish. The region BC is the polishing plateau where the desired smooth surface finish is developed. During the electropolishing process, two films are formed on the surface of the specimen. The first film that is closest to the specimen is a thin solid film (~1 to 10 nm thick) that is constantly being dissolved and reformed. The second is a micron thick viscous film that contains a high concentration of ions and is responsible for the smoothing process. This viscous film is often visible in clear electrolytes as a coloured trail falling from the specimen, as shown in Fig. 2.3b. The majority of specimens are polished in this plateau region. At higher voltages (CD), the polishing process continues and there is a large increase in the current density that is partially due to gas evolution at the

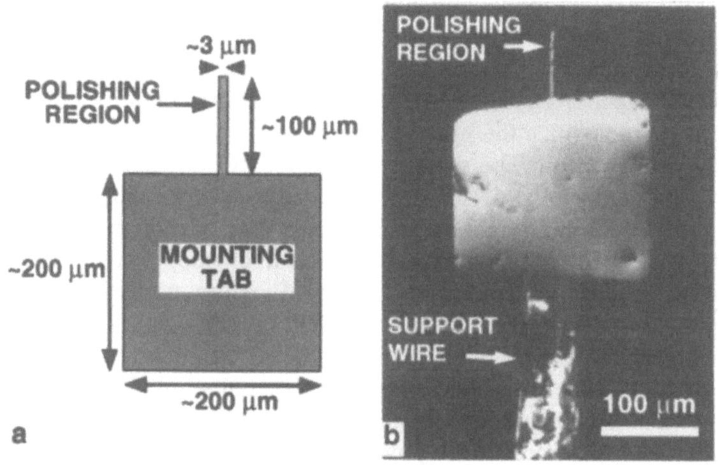

Fig. 2.2. a) A typical lithography mask for producing field ion specimens from thin films. The field ion specimen is fabricated from the narrow section protruding from the square base. b) A typical specimen blank fabricated by lithography methods that is bonded to a support wire,Courtesy D. J. Larson, University of Oxford [13].

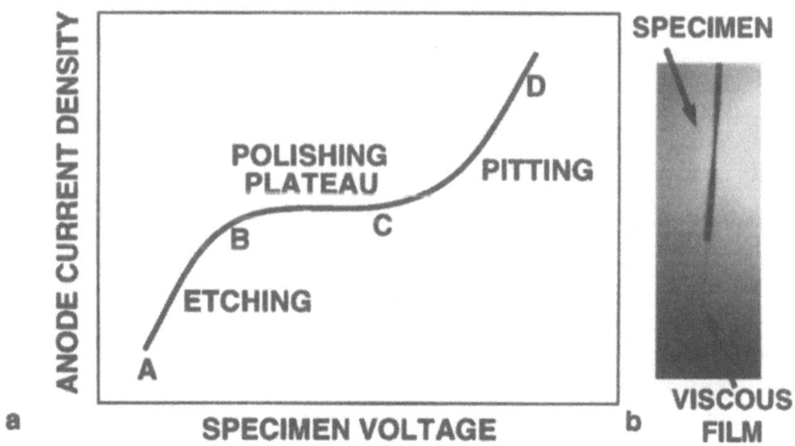

Fig. 2.3. a) The voltage/current density curve for a typical electropolishing solution. Electropolishing is typically performed within the plateau (BC) region of the curve. b) Example of the semitransparent viscous film produced during electropolishing of a stainless steel specimen.

anode. This gas evolution locally disrupts the viscous film and can lead to the formation of pits on the surface.

A list of suitable electrolytes and the conditions for a variety of materials is given in Table 2.1. These electrolytes and polishing conditions should only be used as a guide since the precise conditions vary among laboratories and with alloy composition. The methods listed for iron for the first and second stages represent a general purpose or universal electrolyte that can be applied to many different materials. These electrolytes are often derived from solutions developed for transmission electron microscopy. Additives such as glycerol or 2-butoxyethanol (butyl cellosolve) are used to increase the viscosity of the electrolyte and to change the characteristics of the viscous layer. Temperature can be used to control the reactivity and viscosity of the electrolyte and hence the speed and quality of polishing. The solutions are sometimes cooled to quickly dissipate the heat generated by exothermic electrochemical reactions or the power supplied to the cell. Some materials are electropolished with the use of molten salts at elevated temperatures.

In materials that contain several phases of different chemical compositions, it is desirable to select electropolishing conditions that produce a similar rate of removal of material for all phases present. In practice, it is not always possible to achieve this goal and the resulting specimen may exhibit phases that protrude from the surface. Unequal rates of electropolishing may also be observed at grain or other types of boundaries and this can lead to grooves on the surface of the specimen. Another artifact that may be produced by electropolishing is the formation of thin surface films, as shown for a copper-containing aluminum alloy in Fig. 2.4. In this example, copper dissolved in the electropolishing solution has been redeposited onto the surface of the specimen. These thin surface layers may generally be removed by field evaporation (§1.2.2).

2.3.1 Double layer technique

The normal method of electropolishing is a two-stage process. In the first stage, the blank is suspended in a 5 to 7-mm-thick layer of electrolyte. The top of the blank is positioned in air and the bottom of the blank is in a dense inert liquid such as a perfluorinated polyether to prevent polishing, as shown schematically in Fig. 2.5. A counter electrode, normally made of a noble metal, such as gold or platinum, is positioned a short distance from the specimen. In many cases, the size and shape of the counter electrode is not critical and a simple wire electrode may be used. In other cases, a circular cathode that is coaxial and equidistant from the specimen is required. When voltage is applied between the specimen and the counter electrode, only the central section of the specimen in the electrolyte is polished and a necked region is gradually formed. In some cases, the specimen has to be gently

Table 2.1. Standard electropolishing recipes.
The notations 1st and 2nd refer to the first and second stages of a two stage procedure and a and b are alternative electrolytes. Electropolishing is performed at room temperature unless otherwise stated. Appropriate safety precautions should be taken before preparation and use of these chemicals due to the toxic, corrosive, flammable, explosive and otherwise harmful nature of some of these mixtures.

Material	Electrolyte	Conditions
Aluminum alloys	a) 1-10% perchloric acid in methanol b) 25-30% HNO_3 in methanol c) 80% HNO_3 in water	5-10 V AC: –10°C 5-7 V DC: –30°C 3 V AC: 0°C
Boron	KOH or $NaNO_3$	10-20 V AC: molten
Chromium	as iron	as iron
Cobalt alloys	1st) 25% chromic acid in water 2nd) 10% HCl in water	6 V AC 3.2 V AC: stainless steel counter electrode
Copper alloys	a) 1st) 70% orthophosphoric acid in water 2nd) 100% orthophosphoric acid b) 10 g $Na_2CrO_4.4H_20$ in 100 ml acetic acid	1-5 V AC or 16 V DC: copper counter electrode 8-15 V DC
Gold	20 wt % KCN in water	4-5 V DC
Iridium	30% chromic acid in water	~10 V AC
Iron alloys and steels	1st) 25% perchloric acid (70%) in glacial acetic acid 2nd) 2% perchloric acid in 2-butoxyethanol	10-25 V DC 10-25 V DC
Manganese alloys	as iron	as iron
Molybdenum	a) 5M NaOH in water b) 12% H_2SO_4 in water	6 V AC 6 V DC

Nickel alloys	1st) 10% perchloric acid (70%), 20% glycerol in ethanol 2nd) 2% perchloric acid in 2-butoxyethanol	22 V DC 25 V DC
Niobium	10% HF in HNO_3	1-3 V DC
Platinum	80% $NaNO_3$, 20% NaCl	3-5 V DC: molten
Rhenium	50% H_3PO_4 in H_2O_2 (30% weight per unit volume)	3-9 V DC
Ruthenium	25% KCl or KOH in water	1.5–30 V AC
Tantalum	45% HF, 22% H_2SO_4 22% H_3PO_4 in acetic acid	15 V DC: Platinum counter electrode
Titanium	6% perchloric acid, 34% n-butanol in methanol (circulating electrolyte)	50-60 V DC: -50°C
Tungsten	5% NaOH in water	1-5 V AC
Vanadium	15% H_2SO_4 in methanol	6.5 V DC
Zirconium and Uranium alloys	as iron	as iron

raised and lowered in the electrolyte to alleviate the effects of preferential attack at the air-electrolyte interface. This up and down motion may also be used to polish selective regions of the specimen to improve the taper angle. The electrolyte, applied voltage and temperature of the solution are chosen to remove material rapidly and uniformly from the specimen. Either direct or alternating currents may be used depending on the material. In the case of direct current, the specimen (anode) is attached to the positive side of the power supply. If alternating voltage is used, the shape of the wave, phase angle, frequency may be adjusted to optimize the polishing conditions. Care is required to maintain the temperature of the solution due to the heat of reaction that is generated by the electropolishing. Best results are usually obtained with the use of freshly prepared electrolytes. In some materials such as fine tungsten wire, the electropolishing can be continued until the weight

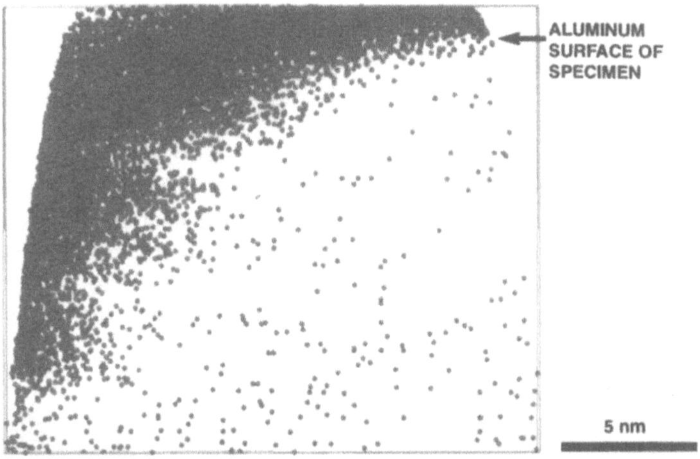

Fig. 2.4. A copper atom map illustrating the presence of an ~5-nm-thick copper film formed on the surface of a copper-containing aluminum alloy after electropolishing. From a collaboration with F. Danoix, University of Rouen, France.

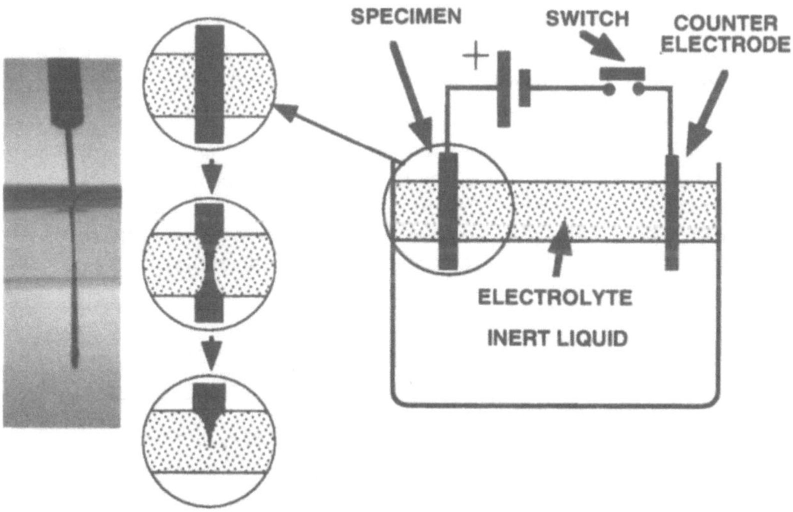

Fig. 2.5. Specimen preparation technique for the double layer method. Electropolishing only occurs in the central section of the specimen in the layer of electrolyte and produces a necked region.

of the lower half of the specimen is too large to be supported by the necked region and the lower half separates and electropolishing is terminated. However, in most materials, electropolishing is stopped just before separation and the specimen is transferred to a second stage with a more controllable electrolyte for the final stages of polishing. This second stage is usually performed in a simple bath of electrolyte rather than with an inert layer. Both halves of the blank may be used as field ion specimens. Some local mechanical damage can occur in the apex region of the needle. If necessary, this damaged region may be removed with the use of the micropolishing method (§2.3.3). Since there is a sudden decrease in the current when the lower half separates, it is possible to electronically monitor polishing current and automatically interrupt the electropolishing.

2.3.2 Chemical etching and the dip method

A related method is chemical etching. In this method, a wire or blank is simply placed in a suitable solution for a given time without any applied voltage [2], as shown in Fig. 2.6. Solutions for selected materials are given in Table 2.2. The rate of material removal from the sides of the blank is balanced by the rate of material removal from the end of the blank. If these rates can be correctly balanced and the starting dimensions of the wire or blank are appropriate, a sharp needle can be formed. Progress has to be frequently monitored in an optical microscope. This method is most applicable when applied to blanks with circular cross sections with small initial diameters. This procedure may also be used with an applied voltage and is then referred to as the "dip" method.

Table 2.2. Chemical etching solutions for selected materials. Appropriate safety precautions should be taken before preparation and use of these chemicals due to the toxic, corrosive, flammable, explosive and otherwise harmful nature of some of these mixtures.

Material	Solution
Gallium Arsenide	44% H_2SO_4, 28% H_2O_2 (30% weight per unit volume of solution) in water
Gallium-Indium Arsenide	0.5% H_2O_2, 0.5% H_2SO_4 in water
Gallium Phosphide	25% HNO_3 in HCl
Nb_3Sn	10% HF in HNO_3
Silicon	30-50% HF in HNO_3

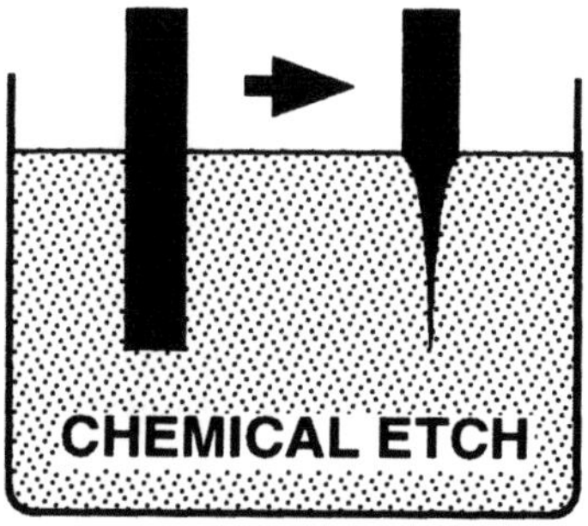

Fig. 2.6. Chemical etching method in which the rate of removal of material from the sides of the specimen is balanced against the rate of removal from the end of the specimen.

2.3.3 Micropolishing

Another method that is commonly used to sharpen blunt or damaged specimens is micropolishing [16]. In this method, the specimen is positioned so that the end of the needle is located in the center of a drop of electrolyte suspended in a wire loop, as shown in Fig. 2.7. The loop is normally made of platinum or another noble metal and is typically ~3 mm in diameter. A voltage is then applied between the specimen (anode) and the wire loop (cathode) to electropolish the portion of the specimen in the electrolyte. The drop of electrolyte has to be frequently refreshed due to its small volume. By moving the specimen with respect to the wire loop and intermittently applying the voltage at appropriate times, material may be selectively removed from a desired region of the specimen. This operation is usually performed under a low power (~30X) optical microscope so that the precise alignment between the specimen and the loop, the region that is being polished and the rate of polishing can be monitored.

2.3.4 Pulse polishing

Pulse polishing is usually performed to produce a specimen with a desired microstructural feature in the apex region [17-22]. In this method, the needle-shaped specimen is initially examined in a transmission electron microscope. If a desired feature is observed at a distance along the axis of the needle, the specimen is removed from the transmission electron microscope and transferred to an electropolishing cell. Small quantities of material are then removed by applying one or more short duration voltage pulses to the specimen. The duration of the voltage pulse (typically 1 to 10 ms) and the

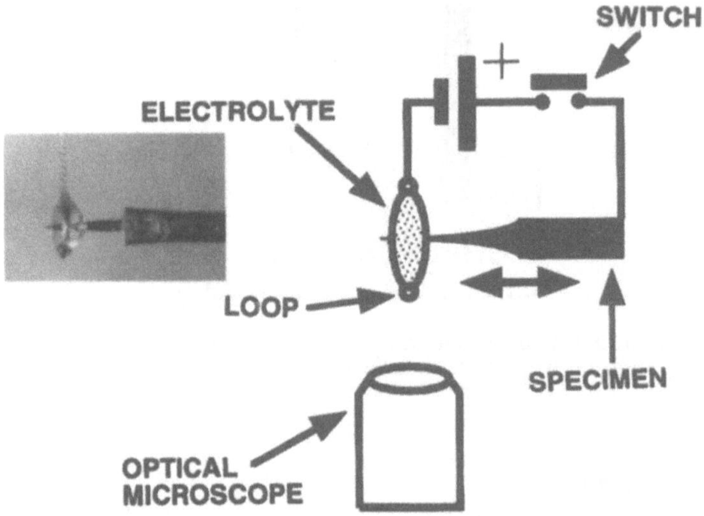

Fig. 2.7. Micropolishing in which the needle-shaped specimen pierces a drop of electrolyte suspended in a wire loop. The operation is monitored with a low magnification optical microscope.

number of pulses that are required to remove a given amount of material can be approximately calibrated for a given material and electrolyte. The duration of the voltage pulse must be sufficiently long to ensure that the solid film and the viscous layer (§2.3) have time to form. The specimen is then re-examined in the transmission electron microscope to determine the amount of material removed and the new position of the feature of interest relative to the apex region of the specimen. This process is continued until the feature is brought into the apex region of the needle where it can be analyzed in the atom probe, as shown in Fig. 2.8. This process enables features that are present in low number densities to be analyzed.

2.4. Milling

If a suitable electropolishing or chemical solution cannot be identified, the normal alternative is ion milling. These methods tend to be significantly slower than electropolishing techniques and can damage the specimen.

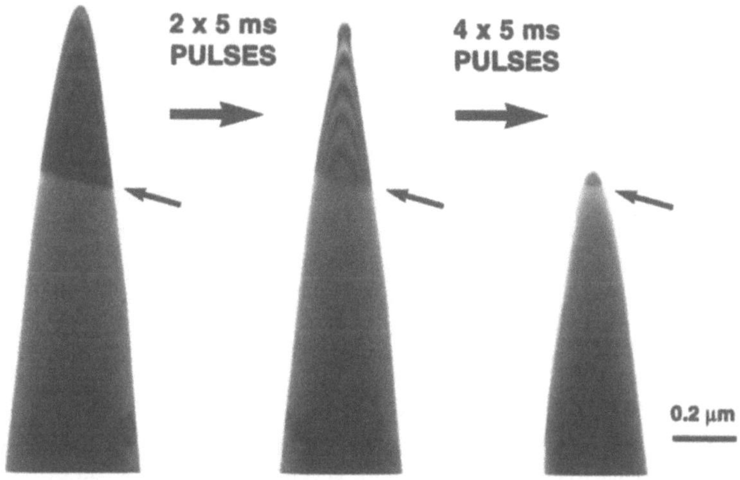

Fig. 2.8. Typical sequence of transmission electron micrographs recorded during pulse polishing of a stainless steel specimen containing a grain boundary as material is removed from the specimen. From a collaboration with T. F. Kelly, University of Wisconsin and D. J. Larson, Oak Ridge National Laboratory.

2.4.1 Standard ion milling

In this method [22,23], the specimen blank is slowly rotated in the center of one or two ion beams until a sharp needle is formed, as shown in Fig. 2.9. The sputtering rate can be controlled by adjustment of the accelerating voltage and beam current. To reduce the time required, the blank is usually thinned to a blunt needle with any of the methods described above or by mechanical means prior to ion milling. The specimen has to be removed periodically to evaluate the progress in an optical microscope or in a transmission or scanning electron microscope. Precision ion milling of a blunt or fractured specimen can also be used to sharpen or sculpt the needle [24,25]. In this method, the ion beam is scanned over a selected region at the end of the specimen and the progress is monitored at intervals during the removal of material.

2.4.2 Sphere-on-surface method

It is possible to fabricate field ion specimens in which the specimen axis is perpendicular to the surface of a film or surface layer. One method is to place some small hard spheres (typically 1-5 μm diameter diamond particles) on the surface of a small section (~1 mm x ~1 mm) of the film and then ion mill

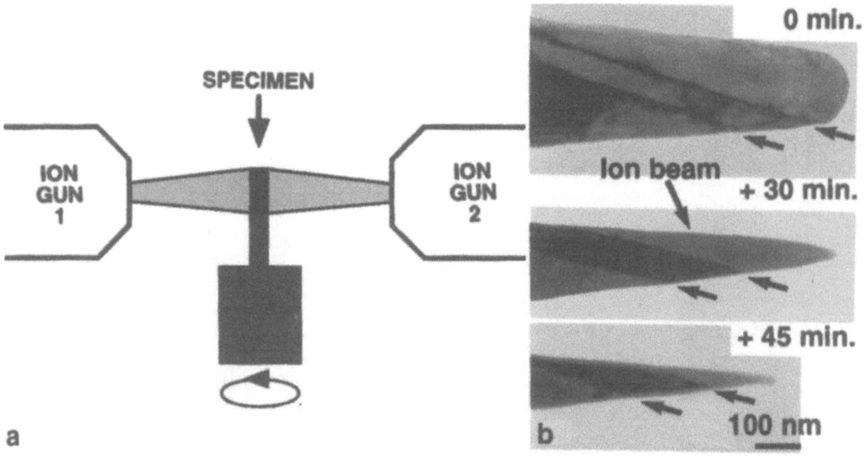

Fig. 2.9. a) A typical ion milling configuration. The blunt needle is slowly rotated in one or two ion beams. b) A titanium aluminide specimen containing a pair of γ-γ interfaces before and after ion milling is shown, courtesy D. J. Larson, Oak Ridge National Laboratory.

normal to the surface [26,27], as shown in Fig. 2.10. Since the spheres mask the surface from the ion milling, a needle is formed under the sphere. Progress has to be checked in a scanning electron microscope at regular intervals to ensure that milling is stopped precisely when the sphere has just been milled away. This method produces several needles at a time and the most suitable one is selected by removing the others. Only limited success has been achieved with this method for field ion microscopy applications since the length of the needle produced is relatively short (~25 μm) [27]. However, this method may find applications for producing specimens for the scanning atom probe.

2.4.3 Focussed ion beam methods

A more successful technique is to mill thin film specimens with a focussed ion beam (FIB) [13-15]. This method permits the progress to be monitored during the milling operation without removing the specimen. To reduce the length of time required in the FIB, a suitable blank is produced by lithography, as described in §2.2. The typical dimensions of the end of the blank that will be fabricated into a needle is ~3 μm wide by the thickness of the film by ~150 μm long. Alternatively, a blunt needle produced by electropolishing or chemical etching may be used as the starting point. Two

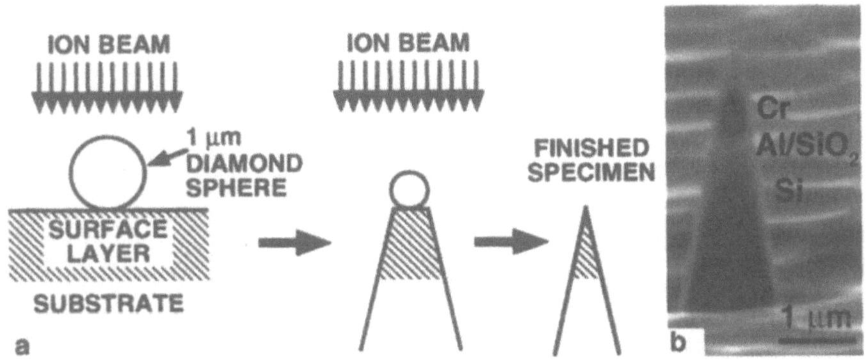

Fig. 2.10. a) Sphere on surface method used to produce field ion needles from the surface layers of a specimen. The diamond particle protects the surface from the ion beam and a sharp needle is formed under the particle. b) An example is shown of a specimen fabricated from a chromium-coated metal (aluminum)-oxide (silica)-semiconductor (silicon) wafer, courtesy D. J. Larson, University of Wisconsin [27].

approaches may be used to sharpen the blank into a sharp needle. In the first, a series of inclined line cuts are made across the end of the blank at equal increments of rotation (e.g., ~90° or ~120°) around the specimen axis, as shown in Fig. 2.11. At the end of the milling process, a pyramid-shaped apex region is formed.

A more desirable circular end form may be produced in the second method. The sample is positioned so that the specimen axis is aligned parallel to and centered on the column of the ion beam. The ion beam is then rastered across the end of the specimen in a circular annular pattern, as shown in Fig. 2.12. This annular method is similar to the action of a pencil sharpener. The milling is generally performed in a series of stages with successively smaller inner and outer diameters and decreasing ion currents. The outer diameter of the initial raster pattern of the ion beam must be larger than the projection of the width of the specimen some tens of microns down the shank in order to prevent the formation of a sharp rim around the specimen. The conditions for a typical three-stage process are given in Table 2.3. The times and the diameters of the annular raster pattern of the ion beam vary from specimen to specimen and with the type of material.

These focussed ion beam methods introduce some artifacts into the specimen. The gallium ions from the liquid metal ion source are implanted into the surface layers of the specimen. The amount of implanted gallium may be minimized by keeping the time spent imaging to a minimum and by

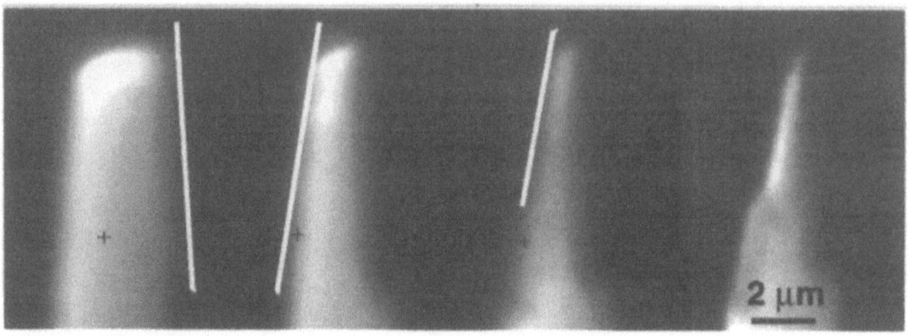

Fig. 2.11. A series of linear cuts are made with a focussed ion beam across the apex region of a copper-cobalt multilayer film to produce a pyramid-shaped specimen. These low resolution images were produced by the secondary electrons produced by the gallium ion beam. Courtesy D. J. Larson, University of Oxford [14].

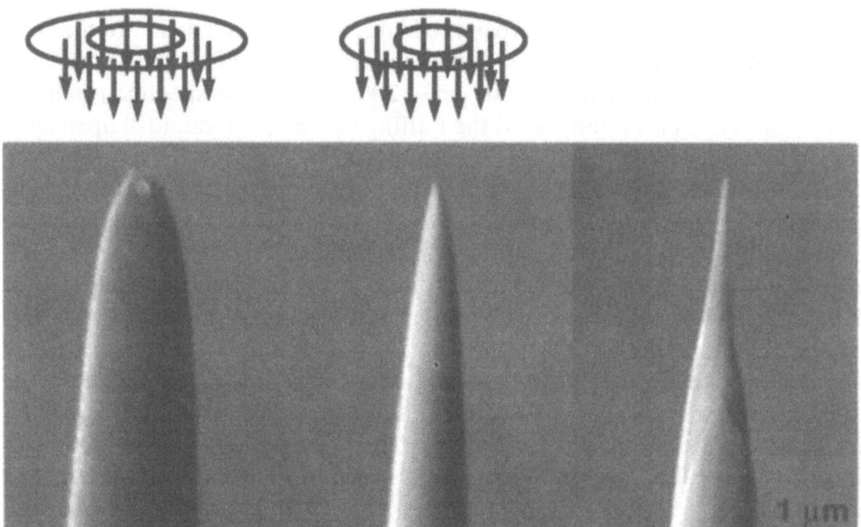

Fig. 2.12. Annular milling with a focussed ion beam. The inner and outer diameters of the annulus are reduced for successive stages of milling. Courtesy D. J. Larson, Oak Ridge National Laboratory.

the use of low currents (typically <50 pA) during imaging. In addition, a zone of amorphous material and some intermixing of the phases or layers present may be produced. These artifacts may normally be eliminated by field evaporation of the surface layers prior to analysis and by analyzing only the central region of the specimen.

Table 2.3. Parameters for a typical three stage annular milling process.

Ion Energy keV	Ion Current pA	Annular Diameter μm		Milling Time s
		Outer	Inner	
30	3000	10	2	180
30	150	3	0.3	45
30	70	1	0.2	20

2.5 Other Methods

The procedures described in the previous sections are by far the most common methods for producing suitable field ion specimens. However, some other methods have been developed for selective materials. For example, field ion specimens of carbon whiskers and silicon carbide fibers have been made by heating the whisker in an oxygen-rich flame.

Field ion specimens may be sharpened in the field ion microscope by reversing the polarity of the voltage on the specimen and producing field emission in the presence of the image gas. These electrons ionize the image gas atoms which are then attracted to the negatively charged specimen where they impact and sputter the surface. The sputtering produces some sharpening as the ions preferentially strike the shank of the needle rather than the apex region. This method is slow and usually results in only a small improvement.

The method of sharp shards simply selects a needle from broken pieces of a brittle material such as high temperature superconductors [28]. These shards are then mounted on wires. This method is normally performed under a low power optical microscope. However, the resulting specimens generally contain cracks or other anomalies resulting from the fracture process.

2.6 Cleaning and Inspection

Following all the specimen preparation methods described in this chapter, the specimen must be thoroughly cleaned to remove all traces of the electrolytes and other contamination. Cleaning is usually performed by carefully rinsing the specimen in water, acetone or methanol. Due to the small dimensions of the apex region of the needle, care must be exercised since the surface tension of a liquid can be sufficient to bend or destroy the needle. Typical methods are to gently pour a stream of cleaning solution on the support in order to allow it to flow down the needle and off the apex region. Alternatively, the specimen may be gently moved in a bath of cleaning solution. Entry and exit into the cleaning bath should always be performed with the specimen axis normal to the air-liquid interface and with the blunt end of the needle intersecting the meniscus first. Vigorous shaking of the needle should also be avoided. Precautions should also be taken during the handling of field ion specimens to prevent static electricity from discharging at the apex of the specimen.

After cleaning, the specimen is dried and checked for suitability in an optical microscope. Typical magnifications that are used are 200 to 500 times. The higher magnifications that are available with the use of oil immersion techniques are not used due to the problem of surface tension and contamination. Although the optical microscope does not have sufficient resolution to resolve the apex region of the needle, the quality of the surface finish and the presence of protruding phases or grooves can be determined. In addition, the sharpness of the needle can be estimated from the taper angle and the presence of interference fringes in the apex region.

Many specimens must be placed into a vacuum immediately after cleaning to prevent oxidation or corrosion of the surface. In some cases, a short period in an ion milling system may be necessary to remove surface films. These surface films may be formed on specimens that have been exposed to air or from artifacts during specimen preparation, such as redeposition.

2.7 Prescreening in Transmission Electron Microscope

Examination of field ion specimens in transmission electron microscopes is a common procedure. This type of examination is possible since the apex region of the needle is sufficiently thin to be electron transparent, as shown in Fig. 2.1. It is advantageous to use a transmission electron microscope with a relatively high operating voltage (e.g., 300 keV) as larger diameter regions of the needle may be penetrated at the higher voltages. This penetration also enables regions further down the shank of the needle to be examined, as shown in Fig. 2.13 for a pressure vessel steel specimen [29]. Most field ion specimen holders that are available for the transmission electron microscopes generally only offer a single rotation about the specimen axis. A second axis of rotation if available is extremely limited due to the typical length of the field ion specimen and the available space between the pole pieces of lens in the electron microscope. The varying thickness and circular cross section of the needle also imposes some limitations on the type of characterization that can be performed in the electron microscope.

Fig. 2.13. Prescreening a pressure vessel steel field ion specimen in a transmission electron microscope to determine the features present in the needle. Courtesy M. G. Burke, U. S. Steel Corp. [29].

Transmission electron microscopy of field ion specimens is often used to assist in the fabrication of a specimen with a particular microstructural feature, such as a grain boundary or a precipitate, in the analyzable apex region of the needle, as discussed in §2.2.4. Transmission and scanning electron microscopy are also used to examine the surface of the specimen in order to determine the quality of the specimen preparation techniques.

Specimens are also examined in the transmission electron microscope at the end or sometimes during an interruption of an atom probe experiment in order to determine the radius of curvature of the apex of the needle at a known applied voltage. This information is required in the algorithm used to reconstruct the three-dimensional data sets so that the correct magnification factors are used (§5.5).

References

1. J. A. Horton and M. K. Miller, *J. de Phys.*, **47-C2** (1986) 209.
2. E. W. Müller and T. T. Tsong, Field Ion Microscopy: Principles and Applications, 1969, Elsevier, New York, NY, pp. 109-127.
3. M. K. Miller and G. D. W. Smith, Atom Probe Microanalysis: Principles and Applications to Materials Problems, Materials Research Society, 1989, Pittsburgh, PA, pp. 37-59.
4. A. J. Melmed, *J. Vac. Sci. Technol.*, **B9** (1991) 601.
5. S. J. Savage and F. H. Froes, *J. Met.*, **36** (4) (1984) 20.
6. A. J. Melmed and R. Klein, *J. de Phys.*, **47-C2** (1986) 287.
7. T. Masumoto, I. Ohnaka, A. Inoue and M. Hagiwara, *Scripta Metall.*, **15** (1981) 293.
8. I. Ohnaka, I. Yamauchi, T. Ohmichi, T. Ichiryu, T. Mitsushima and T. Fukusako, Proc. 5th Intl. Conf. Rapidly Quenched Metals, Sept. 3-7, 1984, Wurzburg, Germany, S. Steeb and H. Warlimont, eds., North Holland, New York, NY, (1985), 111.

9. I. Ohnaka, T. Fukusako, T. Ohmichi, T. Masumoto, A. Inoue and M. Hagiwara, Proc. 4th Intl. Conf. Rapidly Quenched Metals, Aug. 24-28, 1981, T. Masumoto and K. Suzuki, eds., Japan Institute of Metals, Sendia, Japan, (1982) 31.
10. J. Liu, L. Arnberg, N. Backstrom and S. Savage, *Mater. Sci. Eng.*, **98** (1988) 21.
11. R. Maringer and C. E. Mobley, Proc. 3rd Intl. Conf. Rapidly Quenched Metals, July 3-7, 1978, Brighton, UK, B. Cantor, ed., The Metals Society, London, UK, **1** (1978), 49.
12. N. Hasegawa, K. Hono, R. Okano, H. Fujimori and T. Sakurai, *Appl. Surf. Sci.*, **67** (1993) 407.
13. D. J. Larson, D. T. Foord, A. K. Petford-Long, T. C. Anthony, I. M. Rozdilsky, A. Cerezo and G .D. W. Smith, *Ultramicroscopy*, **75** (1998) 147.
14. D. J. Larson, D. T. Foord, A. K. Petford-Long, A. Cerezo and G .D. W. Smith, *Nanotechnology*, **10** (1999) 45.
15. D. J. Larson, D. T. Foord, A. K. Petford-Long, H. Liew, M. G. Blamire, A. Cerezo and G. D. W. Smith, *Ultramicroscopy*, **79** (1999) 287.
16. A. J. Melmed and J. J. Carroll, *J. Vac. Sci. Technol.*, **2A** (1984) 1388.
17. J. E. Fasth, B. Loberg and H. Norden, *J. Sci. Instrum.*, **44** (1967) 1044.
18. B. Loberg and H. Norden, *Ark. Fys.*, **39** (1969) 383.
19. J. M. Papazian, *J. Microsc.*, **94** (1971) 63.
20. A. Henjered and H. Norden, *J. Phys. E. Sci. Instrum.*, **16** (1983) 617.
21. U. Rolander, *J. de Phys.*, **47-C7** (1986) 449.
22. J. M. Walls, H. N. Southworth and G. J. Rushton, *Vacuum*, **24** (1974) 475.
23. M. Hellsing, *Mater. Sci. Technol.*, **4** (1988) 824.
24. K. B. Alexander, P. Angelini and M. K. Miller, *J. de Phys.*, **50-C8** (1989) 549.
25. A. R. Waugh. S. Payne, G. M. Worrall and G. D. W. Smith, *J. de Phys.*, **45-C9** (1984) 207.
26. J. A. Liddle, A. Norman, A. Cerezo and C. R. M. Grovenor, *J. de Phys.*, **49-C6** (1988) 509.
27. D. J. Larson, C. Teng, P. P. Camus and T. F. Kelly, *Appl. Surf. Sci.* **87/88** (1995) 446.
28. A. J. Melmed, *J. de Phys.*, **49-C6** (1988) 67.
29. M. G. Burke and S. S. Brenner, *J. de Phys.*, **47-C2** (1986) 239.

Chapter 3

Field Ion Microscopy

Field ion microscopy is an important step that is almost always used at the start of an atom probe experiment to produce an atomically clean specimen with a well developed end form. It is also used to examine and characterize the microstructural features present in the specimen, and to select the initial area on the surface of the specimen for analysis in the atom probe. It is often used during and at the end of an atom probe experiment to check the progress of the experiment and to determine certain experimental parameters. The overall shape of the field ion image can often yield a reasonable estimate of the shape and aspect ratio of the apex of the needle. The applications of field ion microscopy have been reviewed recently [1].

3.1 Imaging Procedure

The experimental procedure to form a field ion image is described in this section. In addition, the multi-step process of image formation, magnification and spatial resolution of the field ion image are discussed.

3.1.1 Experimental process to form a field ion image

After the field ion specimen has been fabricated, cleaned and inserted into the microscope, the next step is to form a field ion image. The complete sequence of operations to form a field ion image is as follows:

a) Fabricate and clean the field ion specimen, as described in the previous chapter. Mount the specimen in suitable carrier, §4.2.1. Examine the needle in optical microscope to check sharpness and suitability for imaging, §2.6. The specimen may also be examined in a transmission electron microscope to determine if a feature of interest is present, §2.7.

b) Place the specimen into the airlock. When the airlock has been evacuated to an appropriate pressure, transfer the specimen into the preparation chamber and from there into the analysis chamber. If necessary, move the specimen to the field ion microscope position.
c) Cool the specimen to a suitable cryogenic temperature (usually 20 to 60K depending on specimen material and image gas). It is normal to transfer the specimen directly onto a precooled stage. Wait for the base pressure to recover from the transfer operation (typically 20 min.).
d) Valve off the main pump to the analysis chamber and admit a suitable image gas to a pressure of between 1 to 5 x 10^{-5} mbar. This procedure assumes the use of an instrument with a static vacuum during imaging. Alternatively, a dynamical pumping procedure may also be used in which the valve to the main pump is only partially closed and the image gas is continuously admitted to the analysis chamber.
e) Activate the microchannel plate and phosphor screen on which the field ion image will be formed.
f) Increase the standing voltage on the specimen until the field ion image is visible.

The voltage required to form the field ion image depends on the initial radius of curvature of the specimen and the type of image gas used. This imaging step may have to be performed in total darkness if a light shielded video camera is not available. In order to produce a clean and fully developed apex region on a new specimen, contamination, artifacts of the specimen preparation process and surface films have to be removed. As thc voltage on the specimen is increased, some isolated unstable spots appear on the phosphor screen. These first imaging spots are generally due to adsorbed and condensed gas molecules on the surface, which may be removed by gradually increasing the applied voltage. As the voltage is increased further, a relatively dimly imaging surface will often appear in materials that are prone to surface oxidation and especially after the specimen has been examined in an electron microscope where a thin layer of hydrocarbon contamination was deposited on the surface. Care has to be exercised in removing this layer by slowly increasing the applied voltage as the underlying specimen is sharper than the oxide or hydrocarbon layer and it is possible to apply too high a voltage to the specimen and cause premature fracture (§3.1.4). After that layer is removed, the initial metallic surface will appear. The initial surface often contains some irregularities or damage from the specimen preparation process and a few layers of material normally have to be field evaporated by increasing the field until a uniform end form is developed. The voltage that produces the highest quality field ion image is known as the best image voltage (BIV) and the corresponding field, the best image field (BIF). These parameters usually have a very limited range for single phase materials but may extend over several kilovolts in multi-phase materials and alloys.

The high field necessary for field ion imaging and field evaporation also produces a mechanical stress on the specimen. The magnitude of this stress is proportional to the square of the applied field and can reach levels that are close to the ultimate strength of the material. This stress produces elastic dilations in the apex region of the specimen of the order of 1-2%. Therefore, the minimum voltage required to field evaporate a specimen is generally used.

In order to produce a stable field ion image, the field required to ionize the image gas must be slightly lower than the field required to field evaporate the surface atoms. This requirement is usually satisfied by the appropriate selection of the image gas (§4.2.4). Both these fields must be lower than the field that would produce mechanical failure of the specimen. These fields are dependent to different extents on the specimen temperature and material.

After performing an atom probe experiment in the absence of an image gas, it is advisable to turn off the single atom detector and reduce the standing voltage on the specimen prior to imaging. This is advisable because the field required to evaporate ions is generally lower in the presence of an image gas.

3.1.2 Image formation

The formation of a field ion image is a multi-step process, as shown schematically in Fig. 3.1. As the field on the specimen is increased, the image gas atoms in the vicinity of the apex of the needle become polarized and attracted to the specimen. Because of the polarization forces, the gas atoms acquire kinetic energies that are significantly higher than k_BT. For example, the ionization probability of a helium atom with the full polarization energy approaching a metal surface at a best image field of 45 Vnm^{-1} is only ~0.01. Therefore, most of these atoms rebound from the surface without being ionized. The remaining atoms become thermally accommodated to the cryogenic temperature. This process occurs by a series of collisions during which a small amount of kinetic energy is lost each time the atom strikes the surface and rebounds. The overall rate of ionization depends on the time spent in the critical ionization zone, which increases as thermal accommodation progresses. Typically, the ionization probability increases to 0.07 if the helium atom is accommodated to a specimen temperature of 80K and 0.14 at 20K. If the field is high enough (~37 Vnm^{-1} for neon), the image gas atoms on the surface of the specimen can be field ionized by an electron tunneling process (§3.4). These ions are repelled towards the phosphor screen where they produce a spot of light. This process occurs over the entire apex region of the specimen. The resulting distribution of light on the phosphor screen is the field ion image.

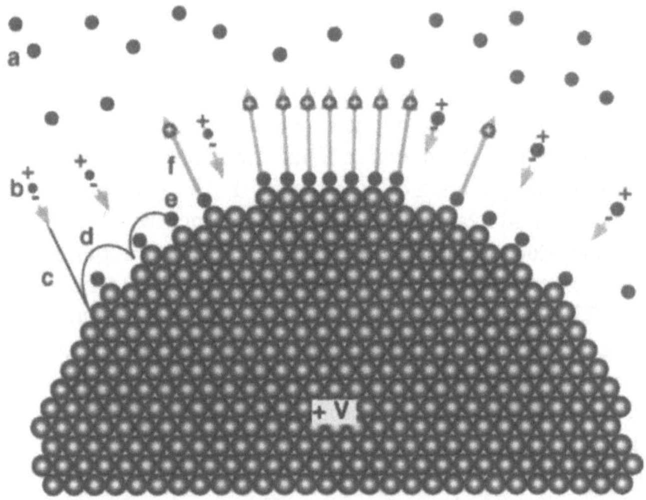

Fig. 3.1. Field ion image formation. When image gas atoms (a) approach the surface of the positively biased specimen, they become polarized (b). These atoms are then attracted (c) to the specimen where they make a series of collisions (d) before they become thermally accomodated (e) to the cryogenic temperature of the specimen. When the atoms are ionized, they are radially repelled from the specimen (f) towards the imaging screen.

3.1.3 Magnification and spatial resolution

The magnification of a field ion image is given by Eqn. 1.6. The magnification depends on the sharpness of the specimen and the specimen-to-detector distance, as discussed in §1.2.4. It has been shown that the field ion microscope is capable of resolving individual atoms on close packed planes, as shown in Fig. 1.1.

The spatial resolution of the field ion image may be expressed in terms of a parameter δ_s, which is defined as the diameter of the image spot produced on the microscope screen by a single atom on the specimen surface, divided by the magnification of the field ion image. The parameter δ_s may be expressed as [2,3]

$$\delta_s = \sqrt{\delta_o^2 + \frac{2\xi h r \sqrt{k}}{\pi\sqrt{2meV}} + 16k\left(\frac{\xi^2 r^2}{eV}\right)\varepsilon_T} \qquad 3.1$$

where δ_O is the diameter of the ionization zone above a given atom, m is the mass of the image gas ion, k is the geometrical field factor, ξ is the image compression factor and ε_T is the transverse thermal energy associated with the image gas atom when it is field ionized. The parameter δ_s is a function of a number of variables, including the diameter of the region over which ionization occurs above a surface atom, the uncertainty in the tangential momentum of the image gas ions, and the lateral thermal velocity of these ions. The best spatial resolution (~0.2 nm) is obtained at low specimen temperatures (10-30 K) with small specimen radii (< 20 nm) and with helium or neon as the image gas.

The precision with which an atom can be located on the surface of a plane where the individual atoms are clearly resolved is significantly better than the spatial resolution. In this case, the position is defined as the centroid of the ion intensity distribution arising from the field ionization process above an individual atom. The location of atoms can normally be measured to better than ± 0.05 nm in surface studies. The precision of the measurement can be improved to ± 0.03 nm with image processing.

3.2 Field Ion Micrographs

In this section, the appearance of many different types of microstructural features in the field ion image is discussed. In addition, some of the microstructural parameters that may be estimated from individual field ion micrographs or from field evaporation sequences are presented.

3.2.1 Single crystals

Some examples of field ion micrographs of single crystals of pure elements are shown in Fig. 3.2. Each of these micrographs exhibits many concentric sets of rings arranged in a geometrical pattern. These rings are a characteristic of field ion micrographs of crystalline materials. The origins of these rings can be visualized by examining a ball model representation of the apex region of a sharp needle, as shown in Fig. 3.3. In this model, the surface atoms that protrude from the mean surface have been highlighted. It is apparent that these protruding atoms are responsible for the rings in the field ion image. Each set of circular rings is due to the intersection of the atomic terraces of a particular crystal plane with the hemispherical surface of the specimen. The relative intensity of the image increases from the center of a pole to the edge of the ring. As the image intensity is proportional to the rate of field ionization of the image gas ions, the distribution of the image intensity indicates that the effective field on the surface atoms increases as the edge of the ring is approached.

The planes can be identified from their position in the field ion micrograph. An example of an indexed field ion micrograph is shown in Fig. 3.4. The typical field of view of a field ion micrograph is ~100°. To assign Miller indices to the planes and their associated poles in a field ion micrograph, the major planes are identified with the use of the relative-prominence rule. This rule states that the planes that exhibit the largest step heights (or interplanar spacings) are the most prominent in the field ion image.

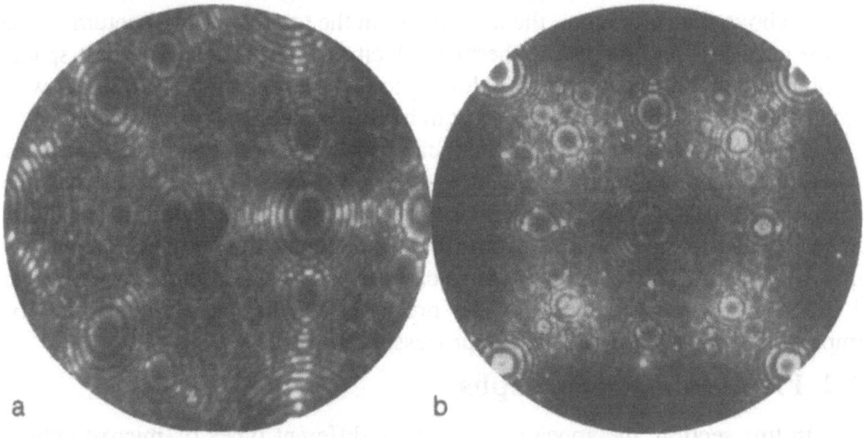

Fig. 3.2. Field ion micrographs of pure elements which exhibit several sets of concentric rings due to the preferential ionization of the image gas atoms at the edges of the atomic terraces: a) neon image of iridium and b) helium image of aluminum (Courtesy K. Hono, National Research Instiitute for Metals, Tsukuba, Japan).

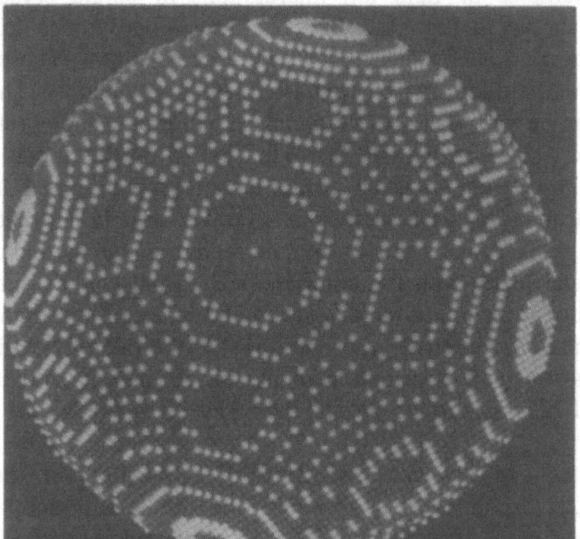

Fig. 3.3. Ball model of the apex region of a field ion specimen with a face centered cubic structure showing the rings formed by the atomic terraces. An (001) plane is at the center and four (111) planes are at the edge of the model.

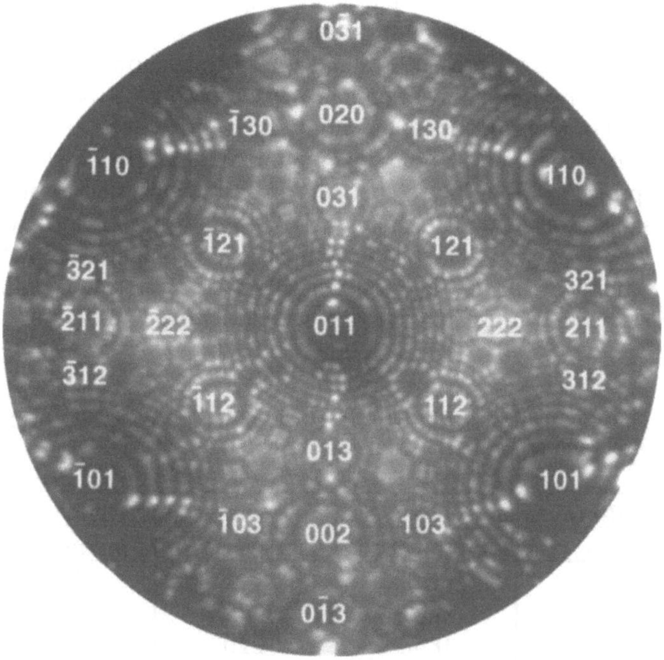

Fig. 3.4. An indexed field ion micrograph of a body centered cubic tungsten specimen.

The relative prominence of the planes in the body centered cubic (bcc) and face centered cubic (fcc) crystal structures are given in Table 3.1. The prominence is defined as the ratio of the interplanar spacing of a specific plane and that of the plane that has the largest interplanar spacing. The Miller indices are assigned based on crystallographic symmetry elements of these major poles. For example in bcc materials, the major 110 and 200 poles have 2-fold and 4-fold rotational symmetry and in fcc materials, the major 111, 200 and 220 poles have 3-fold, 4-fold and 2-fold rotational symmetry, respectively. Once the major poles have been identified, zone lines can be constructed and the other poles may be assigned based on the zone addition rule. This assignment process assumes that the crystal structure of the material is known.

A field ion micrograph is the projection of the surface atoms onto a phosphor screen. In most cases, the phosphor screen is deposited onto a flat surface although in the imaging atom probe the phosphor is deposited onto a spherically curved surface, §1.2.4. Several different types of standard projections such as stereographic, gnomic and orthographic have been used to describe the positions of the poles in the field ion images, as shown in

Fig. 3.5. All these projections match the experimental observations of the positions of the poles at small angles from the center of the field ion image but significant deviations are present at high angles [4], as shown in Fig. 3.6. The stereographic projection is most commonly used because it preserves angular relationships. For all projections, the atoms on well resolved or net planes are not arranged in straight lines in the field ion micrographs but on slightly curved lines. The projection is a key component in the reconstruction of the three-dimensional data collected in the atom probe and is discussed in more detail in §5.5. In more complex crystal structures, stereographic projections and computer simulations of the field ion image (§3.2.10) are often used to assist in the assignment of Miller indices to the crystallographic poles.

Table 3.1. Relative prominences of the major poles for body centered cubic and face centered cubic crystal structures.

Body centered cubic		Face centered cubic	
Plane	Prominence	plane	prominence
110	1.0	111	1.0
200	0.71	200	0.87
112	0.58	220	0.61
130	0.45	113	0.52
222	0.41	240	0.39
231	0.38	224	0.35
240	0.32	244	0.29
244	0.24	260	0.27
226	0.21	460	0.24

Field ion micrographs of pure elements such as tungsten, iridium, rhenium and chromium often exhibit enhanced brightness along particular crystallographic zone lines. This effect is prominent in field ion micrographs of chromium specimens, as shown in Fig. 3.7. This zone line decoration is a result of atoms lying in stable prominent low coordination sites instead of normal surface lattice sites. These zone lines are rarely observed in alloys. In addition, some crystallographic areas of the field image have enhanced regional brightness, as shown in Figs. 3.2 and 3.7. This effect arises from differences in local radii of curvature in these regions.

The atoms on the topmost atomic terrace of a pole are often displaced from their proper position due to surface reconstruction in the presence of the high field. It is unlikely that any reliable information about the local arrangement of the different types of atoms or their bond lengths can be determined directly

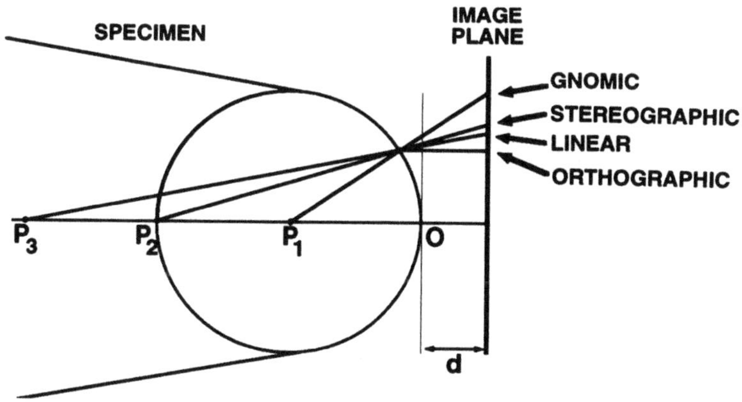

Fig. 3.5. Origins of the stereographic, gnomic, linear, and orthographic projections.

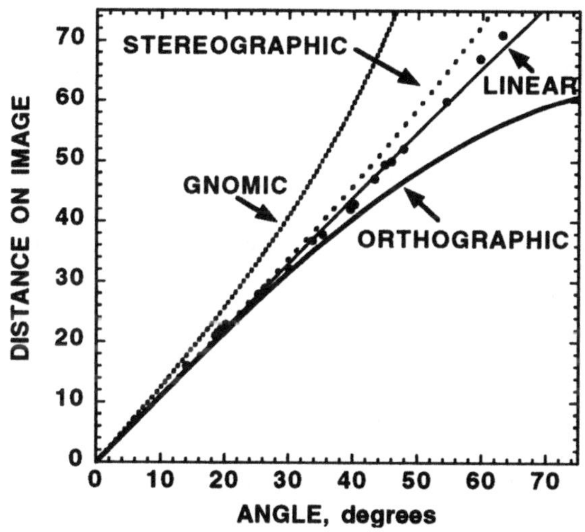

Fig. 3.6. Comparison of the stereographic, gnomic, linear, and orthographic projections with measurements taken from a field ion micrograph of tungsten (after Wilkes et al. [4]).

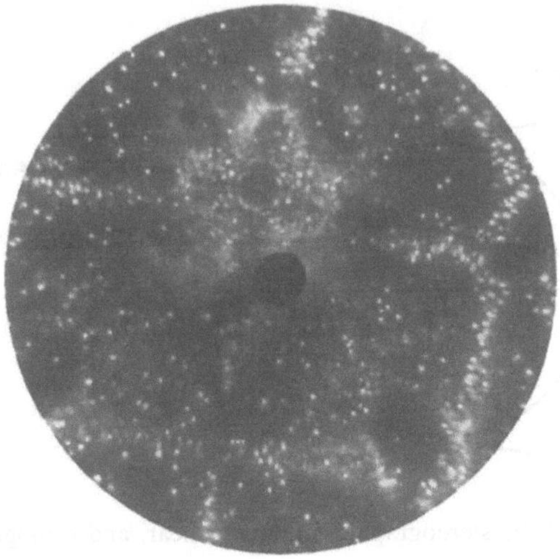

Fig. 3.7. Field ion micrograph of a silicon-doped chromium specimen showing enhanced brightness along the zone lines.

from field ion micrographs due to atomic scale surface reconstruction in the presence of the high field.

3.2.2 Point and line defects

Individual vacancies can sometimes be detected on well-defined planes in pure metals when all the atoms can be resolved. The presence of a vacancy is inferred by the absence of an atom. However, it is common that many vacancies are simply formed by the field evaporation of an atom from within the plane and are therefore artifacts of the imaging process. In special cases, interstitial atoms may also be detected as a bright spot in the appropriate site on a well-resolved plane. However, most of these features are not representative of the material and arise through surface reconstruction or surface diffusion. These factors combined with spatial resolution limitations make it impractical to characterize these types of features in a three-dimensional atom probe.

Larger agglomerations of vacancies, such as microvoids, bubbles, voids or pores, often appear as either dark regions or very bright and streaked regions depending on their size. An example of a brightly-imaging pore is shown in Fig. 3.8. Dark contrast arises because there are no atoms on the surface for the ionization of the image gas. Bright contrast arises when the edge of the void has an extremely small radius of curvature and therefore is a preferential ionization site for the image gas atoms. A void or pore may sometimes be detected a few planes before it intersects the surface of the field

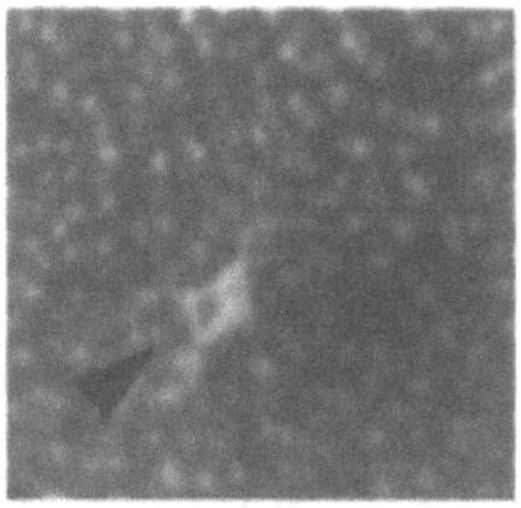

Fig. 3.8. Field ion micrograph of a small brightly-imaging pore in a nickel-base superalloy. The bright contrast arises from preferential ionization of the image gas at the sharp edges of the pore.

ion needle in cases where the gas pressure within the void is sufficient to raise the thin surface.

Dislocations are evident in field ion micrographs when they emerge from the surface in the vicinity of a crystallographic pole [5]. A perfect dislocation with Burgers vector **b** and line **l** will turn the circular rings of planes of normal **n** into a spiral ramp providing that $\mathbf{n} \bullet \mathbf{l} \neq 0$ and $\mathbf{n} \bullet \mathbf{b} \neq 0$. The number of spirals observed depends on the Burgers vector of the dislocation and the plane at which it emerges, as shown in Figs. 3.9a and b for a single and double spiral, respectively. The pitch, p, of the spiral of a dislocation with a Burgers vector $\mathbf{b}=uvw$ that emerges on a (hkl) plane is given by

$$p = hu + kv + lw. \qquad 3.2$$

For example in a face centered cubic material, if an $\frac{a}{2}[110]$ dislocation emerges in a (111) pole, it exhibits a single spiral ($p = 1$), whereas if it emerges in a (220) pole, it exhibits a double spiral ($p = 2$). If an $\frac{a}{2}[110]$ dislocation emerges in a (002) pole, it will not be visible ($p = 0$) as the Burgers vector of the dislocation lies parallel to the surface. If the dislocation emerges outside the center of the topmost terrace, it will initiate the spiral at the closest ring. Segregation to dislocations can sometimes be detected by the presence of brightly-imaging spots, as shown in Fig, 3.9c. This type of segregation can occur at the core of the dislocation or be displaced from the core within the stress field associated with the dislocation.

Stacking faults are occasionally observed in field ion images if the fault vector has a component normal to the specimen surface (i.e., similar to the visibility criterion of a dislocation). As the fault vector of a stacking fault is by

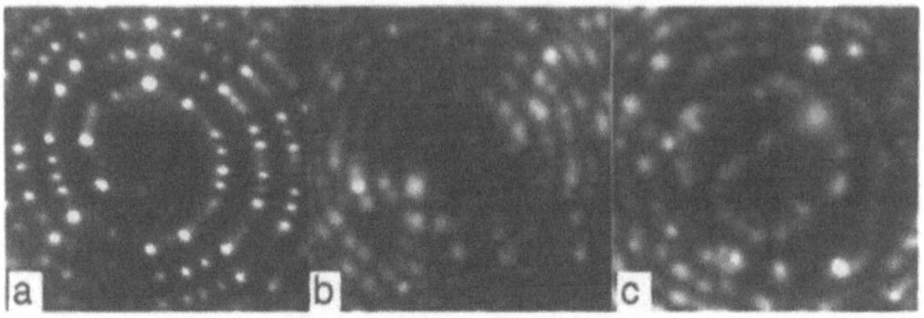

Fig. 3.9. Field ion micrographs of a) a dislocation with a single spiral ($p = 1$) in tungsten and b) multiple spirals in a pressure vessel steel and c) segregation of Mo and P atoms at a dislocation in a pressure vessel steel (from a collaboration with P. Pareige, Oak Ridge National Laboratory).

definition not a repeat distance of the parent lattice, serrated or offset ring configurations are observed.

The stress arising from the high field on the field ion needle can be sufficient to move a dislocation or stacking fault out of the apex region of the needle if the dislocation is not pinned by solute or some other microstructural feature such as a precipitate. Low angle boundaries arising from arrays of dislocations may be similarly observed as a line of dislocations.

3.2.3 Amorphous alloys

Amorphous alloys or glasses by definition do not exhibit any long range order in the arrangement of atoms and therefore the surface of the field ion specimen is a random arrangement of atoms. Consequently, field ion micrographs of amorphous alloys do not exhibit any ring structures, as shown in Fig. 3.10. Although truly amorphous alloys never exhibit ring structures in the field ion micrographs, the absence of rings is not proof of a non-crystalline structure.

3.2.4 Solid solutions

In dilute alloys, it may be possible to distinguish the solute atoms from the matrix atoms by their appearance in the field ion micrographs, as shown in Fig. 3.11. Both bright and dim solute atoms have been documented [1]. Solute atoms that have a higher evaporation field than the matrix are retained on the surface during field evaporation and tend to occupy prominent sites that exhibit enhanced field ionization of the image gas. As the solute content of the alloy increases, the field ion images generally exhibit less regularity than pure metals. In some highly alloyed systems, only the major poles are evident, as shown in a field ion micrograph of a stainless steel in Fig. 3.12a. This effect is due to small displacements of the surface atoms to reduce the local strain caused by the small differences in size of the atoms. This effect will also slightly degrade the spatial resolution of the three-dimensional data because the

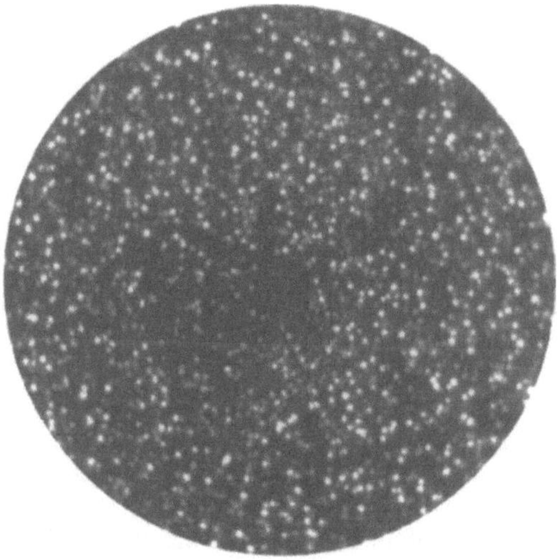

Fig. 3.10. Field ion micrograph of an as-prepared $Pd_{40}Ni_{40}P_{20}$ bulk metallic glass showing the random distribution of spots representative of an amorphous material.

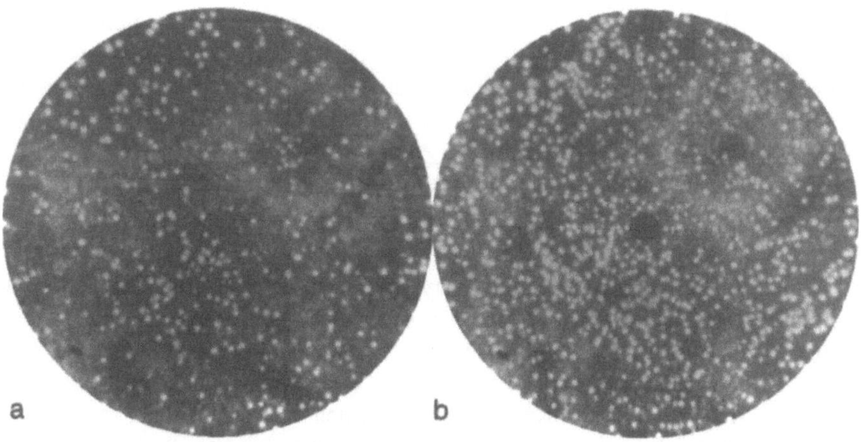

Fig. 3.11. Field ion micrograph of a dilute solid solution in a 4130 steel containing 1% Pt, a) after pulsed field evaporation and b) after DC field evaporation. Platinum atoms are preferentially retained on the surfaces after DC field evaporation.

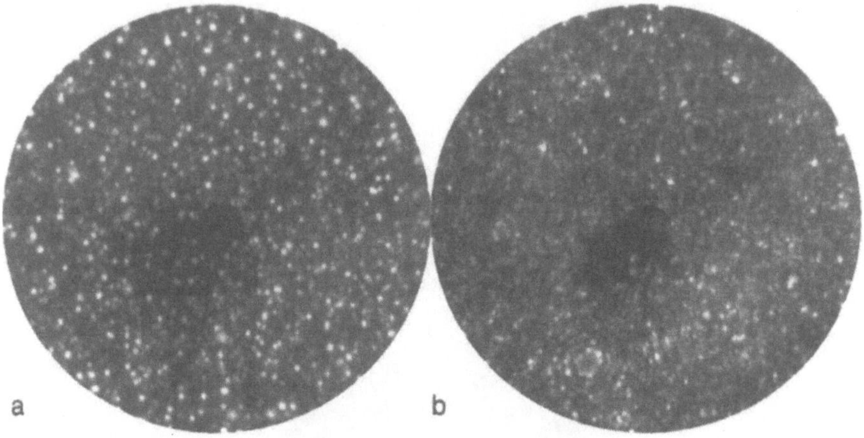

Fig. 3.12. Field ion micrograph of high solute solid solutions: a) a 304 stainless steel in which no major crystallographic poles are evident (from a collaboration with E. A. Kenik, Oak Ridge National Laboratory) and b) an equiatomic AgPd alloy.

atoms are displaced from their normal lattice sites. Some concentrated solid solutions exhibit good quality field ion images, as shown for an equiatomic silver-palladium alloy in Fig. 3.12b.

If the specimen is field evaporated by simply increasing the standing voltage on the specimen, the weakly bound atoms may be preferentially evaporated and the strongly bound atoms retained (§5.3). In this case, the surface of an alloy may not be representative of the bulk material. A more representative surface may be obtained by examining the specimen immediately following analysis in the atom probe. Alternatively, pulsed field evaporation with the same parameters (i.e., appropriate pulse fraction and standing voltage) may be used during field ion microscopy.

3.2.5 Clusters and precipitates

Solute clusters and precipitates typically exhibit regions of different contrast in the field ion micrographs. The contrast mechanism is similar to that of single solute atoms. Generally, the higher the melting point of the phase, the higher the sublimation energy, the higher the evaporation field, the greater its protrusion from the mean surface of the specimen and the brighter its image. Examples of brightly-imaging clusters are shown in Fig. 3.13. Selected examples of brightly- and darkly-imaging precipitates are shown in Fig. 3.14. A cluster is simply an aggregation of solute atoms whereas a precipitate is generally regarded as a volume of different composition with a well-defined crystal structure and interface. However, at extremely small sizes (< 2 nm), it

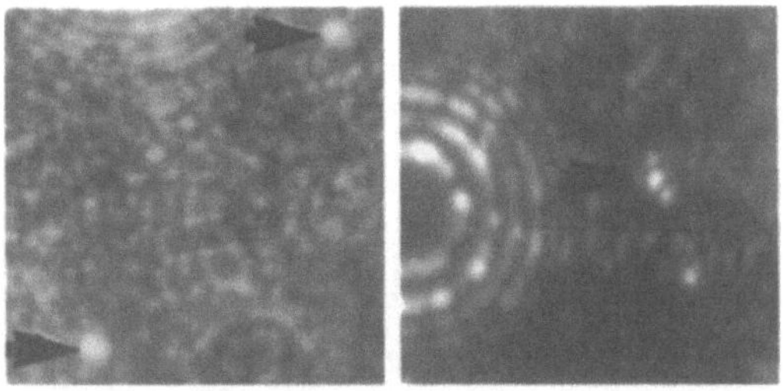

Fig. 3.13. Field ion micrograph of brightly-imaging clusters, a) boron clusters in boron-doped Ni_3Al and b) a phosphorus cluster in neutron-irradiated pressure vessel steel.

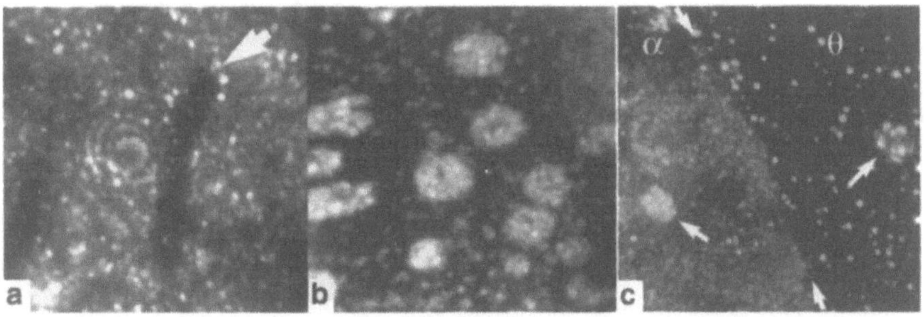

Fig. 3.14. Field ion micrographs of precipitates: a) darkly-imaging cementite (Fe_3C) platelet in α-iron, b) spherical brightly-imaging G-phase precipitates in a duplex steel and c) ~2 nm-diameter brightly-imaging MC precipitates in both the α-iron and cementite (θ) phases of a low alloy steel.

is often impossible to establish a distinction between a cluster, an embryo, and a precipitate.

In some cases, there is little or no contrast to distinguish a precipitate from the matrix phase under the chosen imaging conditions. However, the contrast can be altered by changing the imaging parameters such as the temperature or the standing voltage on the specimen. An illustration of the temperature effect is shown in Fig. 3.15. At the opposite extreme, it may not be experimentally practical to change the imaging conditions to image the matrix and precipitate simultaneously, when the imaging characteristics of the two phases are markedly different. Such a situation arises for imaging of refractory precipitates in alloys, as shown in Fig. 3.16. In this example, the matrix does not show any image contrast when the imaging conditions are optimized for the precipitate, and vice versa. Some other techniques for detecting precipitates are discussed in §3.2.9.

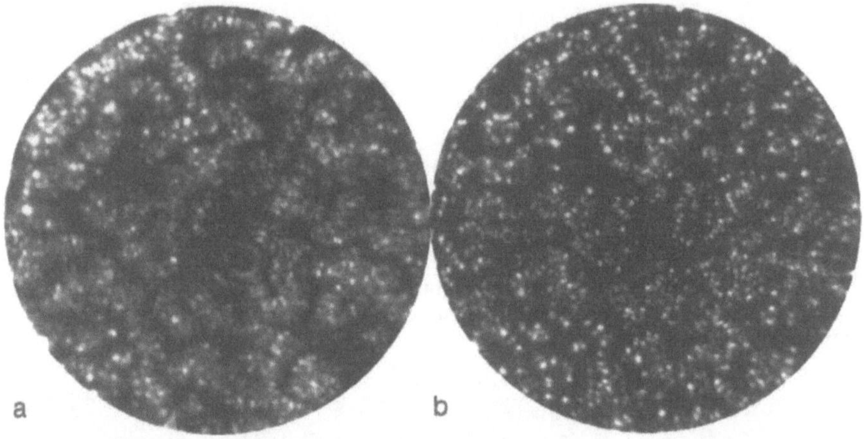

Fig. 3.15. Field ion micrographs of the same Fe-45% Cr specimen acquired at different specimen temperatures, a) 30K and b) 70 K. Note the difference in the contrast between the phases and the size of the individual spots.

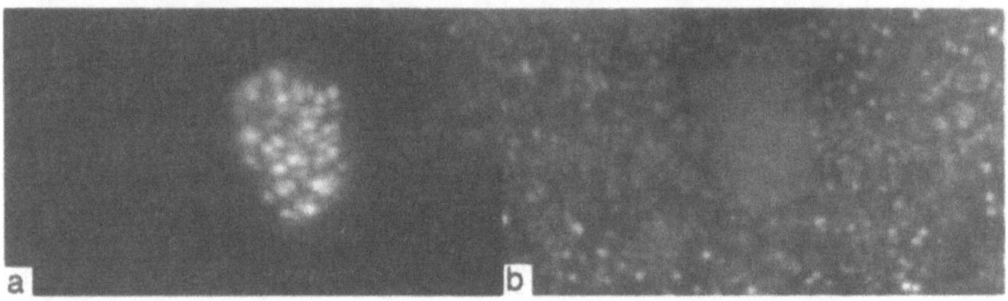

Fig. 3.16. Field ion micrographs acquired at standing voltages selected to optimize the imaging characteristics of a) a V(CN) precipitate and b) the surrounding matrix. No details of the matrix are evident in (a), and the atomic level detail of the precipitate is blurred in (b).

The size of a small precipitate may not be reliably determined from its extent in a field ion image either because its maximum extent may not be evident in an individual micrograph or because there may be large variations in the local magnification between the precipitate and the matrix due to the difference in the local radii of curvatures. The size may be more reliably determined with the persistence size method described in §3.3.4. The morphology of a second phase may also be traced by reconstructing the outline of the phase from the field evaporation sequences. The intersection of the phase with the hemispherical end of the field ion specimen and the image projection must be taken into account in the interpretation of the morphology.

Examples of an interconnected network structure and a triaxially-aligned macrolattice of second phase that form as a result of spinodal decomposition are shown in Fig. 3.17.

3.2.6 Grain and other high angle boundaries

Grain and other types of high angle boundaries such as twin and lath boundaries are evident in field ion micrographs by an abrupt discontinuity in the ring structure on either side of the boundary, as shown in Fig. 3.18. The appearance of the boundary in the field ion image depends on the orientation of the boundary plane with the apex of the specimen. If the boundary plane is exactly parallel to, and passes through, the specimen axis then the boundary plane will appear as a straight line intersecting the center of the micrograph. As the boundary plane gets further from the specimen axis, the appearance of the boundary becomes curved, as shown in Fig. 3.19a. In these special cases, the position of the boundary will not translate during field evaporation. If the boundary plane is normal to the specimen axis, the boundary will appear as a circle in the field ion image. The diameter of the circle will decrease as material is field evaporated, as shown in Fig. 3.19b. In most general cases, the boundary will make some angle to the specimen axis and the boundary will appear curved (i.e., as small circles in crystallographic terminology), as shown in Fig. 3.19c. A method to determine the angle between the normal to the boundary plane and the specimen axis from a field ion micrograph is described in §3.3.5. The direction of curvature of the plane is a useful indicator as to the probable change in position of the boundary in the field ion image as material is field evaporated from the specimen. As a general rule, the position of the boundary in the field ion image moves towards the center of the circle during field evaporation unless the boundary plane is parallel to the specimen axis. This information is useful in positioning the specimen at the start of an atom probe experiment so that the feature of interest will be sampled in the volume of analysis.

Segregation to the plane of the boundary is often evident by brightly-imaging atoms that may merge to form a bright line, as shown in Fig. 3.20. At low coverages, individual brightly-imaging atoms on the boundary may be observed. In some cases, where the solute does not image brightly, such as carbon, a dark line at the boundary may be evident. An approximate estimate of the coverage may be made by counting the number of these brightly-imaging atoms per unit length along the boundary. The coverage may also be estimated from an atom probe analysis (§6.4.6).

The orientation relationship between the grains can be determined from positions of the poles in the field ion micrograph if there is sufficient area visible in each grain to establish the orientation. Additional information on the

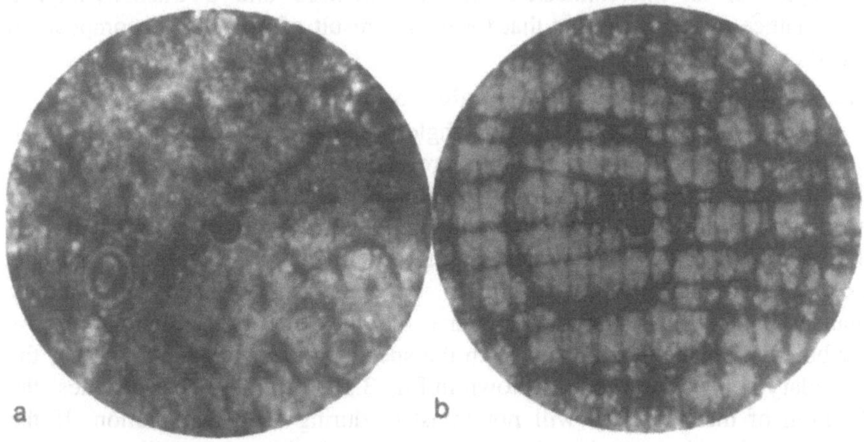

Fig. 3.17. Field ion micrographs of a) an interconnected or percolated network structure of Fe-rich α (bright) and Cr-rich α′ (dark) phases in an Fe-32% Cr alloy aged for 10,000 h at 470 °C and b) a triaxially-aligned macrolattice of Fe-rich α phase particles (bright) in an B2-ordered FeBe phase matrix (dark) in an Fe-25% Be alloy aged for 15 h at 375 °C.

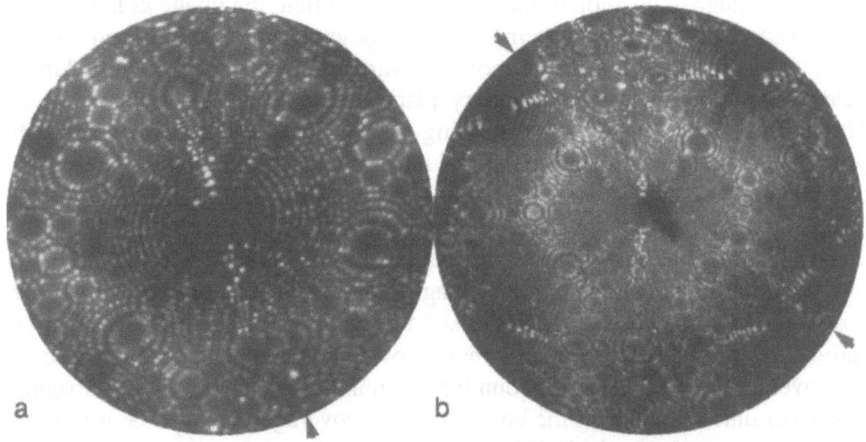

Fig. 3.18. Field ion micrograph of a grain boundary in tungsten. Note the abrupt change in the ring systems either side of the boundary.

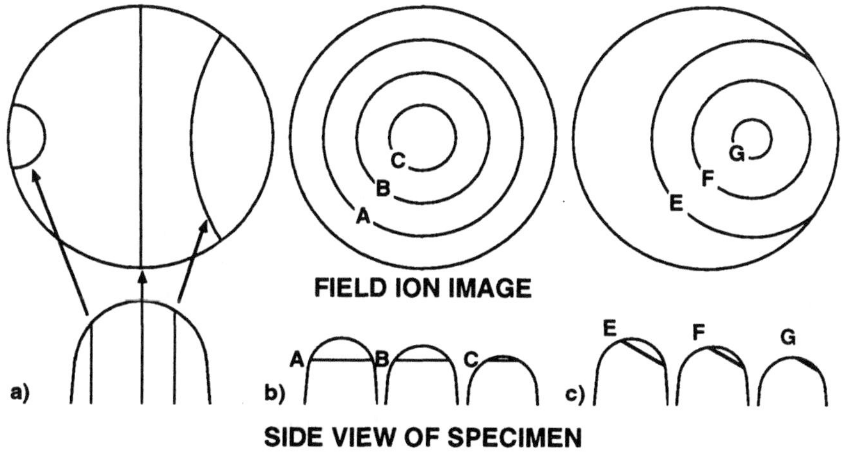

Fig. 3.19. Appearance of grain boundaries in the field ion micrograph. a) boundary parallel to specimen axis, b) boundary perpendicular to specimen axis and c) boundary inclined to specimen axis.

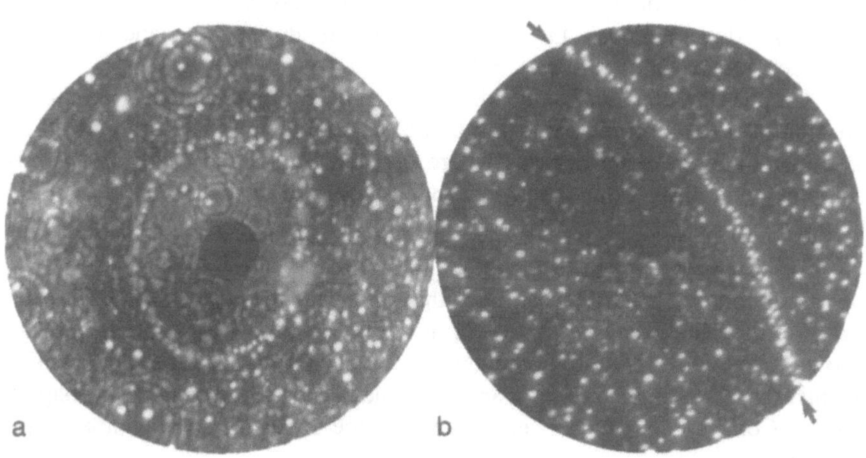

Fig. 3.20. Field ion micrographs of segregation at boundaries: a) a fire resistent steel (from a collaboration with F. Kelly, University of Belfast) and b) a 304 stainless steel (from a collaboration with E. A. Kenik, Oak Ridge National Laboratory).

orientation relationship between the grains may be obtained by examining the field ion specimen in the transmission electron microscope.

3.2.7 Order

Field ion micrographs of long range ordered materials often exhibit well delineated ring structure. Some examples are shown in Fig. 3.21. A characteristic feature of ordered materials is the presence of bright and dim atomic terraces or rings in the field ion image. This contrast arises because of the differences in the ionization rates of the image gas at the ordered planes of different composition. For example, the {001} planes of B2-ordered compounds (such as NiAl) alternate as A and B planes and those in $L1_2$-ordered compounds (such as Ni_3Al) alternate as A and mixed A+B planes. In some cases, the imaging characteristics of the dimmer planes may become so weak as to make these planes invisible. This effect significantly alters the relative prominence of the major poles in the image. For example in a B2-ordered compound, if only one element is imaging, it turns the field ion image into that of a simple cubic compound and changes the prominence from (110), (200), (112), (130) and (222) to (100), (110) (111), (120) and (112). This effect can also effect the accuracy of the radius measurement (§3.3.1).

In some materials that have several different types of layers, such as $YBa_2Cu_3O_{7-\delta}$ high temperature superconductors, one set of planes can dominate the image. This dominance results in circular, curved or almost parallel lines in the image depending on the orientation of the dominant plane with the specimen axis, as shown in Fig. 3.22.

Antiphase boundaries can be identified by the abrupt change from a dimly-imaging plane to a brightly-imaging plane, as shown in a $L1_2$-ordered Ni_3Al specimen in Fig. 3.23. In cases where the unit cell of the ordered phase is complex, antiphase boundaries can be difficult to distinguish from grain boundaries.

3.2.8 Artifacts

Occasionally, bright curved lines that run from the edge of the field ion image to the center may be observed, as shown in Fig. 3.24. These lines are known as comets. It is thought that comets are caused by some feature, such as a whisker, protruding from the shank of the needle outside the normal imaging region. Comets can often be eliminated by the reduction of the image gas pressure.

Another artifact is the presence of bright streaks across the image, as shown in Fig. 3.25. A distortion of the rings and an elongation of the image spots generally accompany streaks. Streaks are a result of a grooved, cracked or fractured specimen. Some less severe grooves may be eliminated by field evaporation. Specimens containing cracks and grooves in the near apex region of the specimen should be avoided because these features could act to

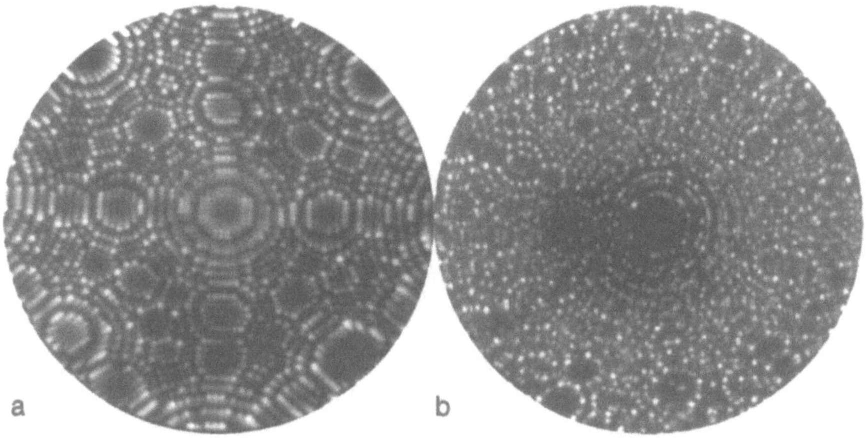

Fig. 3.21. Examples of field ion micrographs of ordered alloys a) Ni_4Mo and b) Ni_7Zr_2.

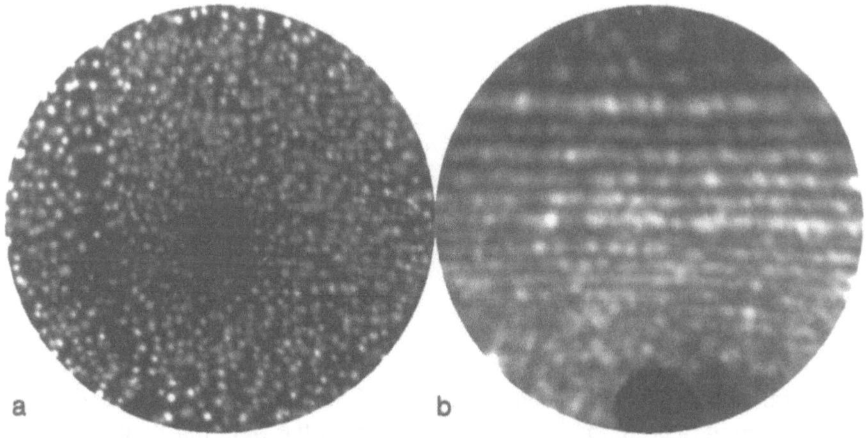

Fig. 3.22. Field ion micrographs of a $YBa_2Cu_3O_{7-\delta}$ high temperature superconductor: a) full image that exhibits some crystallographic poles and b) a magnified region of a field ion image in which the edges of the a-b planes dominate the image. Note the change in constrast of some of the a-b planes, indicative of stacking faults.

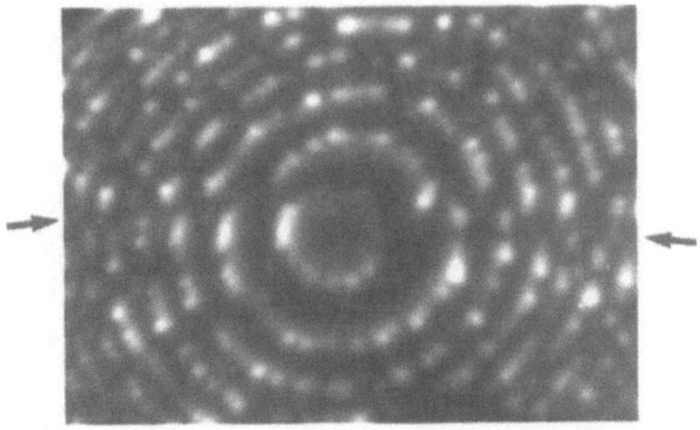

Fig. 3.23. Antiphase boundaries in Ni_3Al. Note the abrupt discontinuity in the brightness of the atomic terraces across the antiphase boundary.

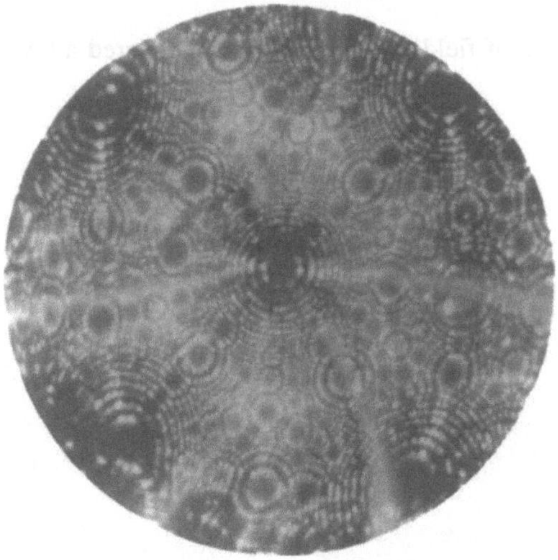

Fig. 3.24. Field ion micrograph of a tungsten specimen that exhibits comets. This artifact can be eliminated by reducing the pressure of the image gas.

concentrate the stress and initiate mechanical failure of the specimen. Care is also required in interpreting features at the edge of the field ion image because they may not be representative of the material. In some materials, it is common for surface films, grooves and other artifacts of the specimen preparation process to remain even after extensive field evaporation.

Some artifacts may arise from the high stress on the specimen due to the applied electric field. Additional stress may be generated in multiphase materials as a result of different coefficients of thermal expansion during the cooling to cryogenic temperatures. The most common manifestation of these

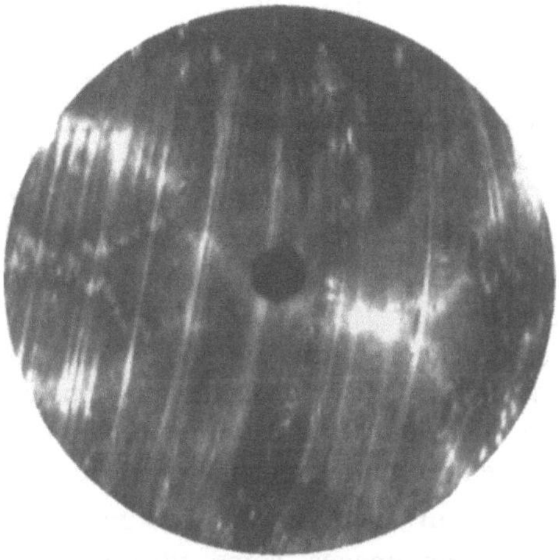

Fig. 3.25. Field ion micrograph of iridium that exhibits streaks across the image due to the presence of grooves at the apex of the specimen.

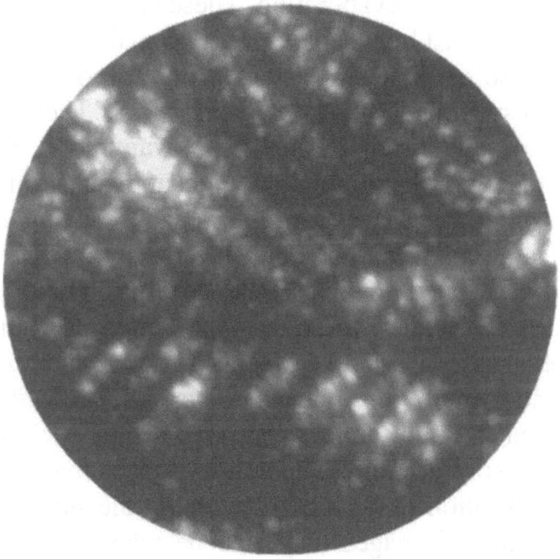

Fig. 3.26. Microtwins formed in a field ion specimen of the Tishomingo meteorite, caused by the field-induced stress on the specimen.

effects is that it can make glissile dislocations migrate out of the apex region of the specimen. It can also initiate failure of the specimen by in-situ microtwinning (§3.2.8). An example of some microtwins produced in a specimen of the Tishomingo meteorite as a result of the applied voltage is shown in Fig. 3.26. In some cases, the effect can be reversible by changing the applied voltage on the specimen. These effects often cause failure of the specimen.

Field ion specimens can fail in several different modes such as the melting of the apex region, cleavage and slip of a portion of the end of the specimen. A typical fracture surface of a broken specimen that exhibits multiple emitting regions is shown in Fig. 3.27. Each of these regions is an individual point that acts as a separate field ion needle. The images from these points can overlap. It is advisable to discard these failed specimens as the microstructure may not be representative of the material.

Etching of the specimen can occur by field-assisted corrosion and results in dark areas in the vicinity of specific crystallographic poles in the images. In severe cases, such as a sudden major leak in the vacuum system, the size of the image can shrink as the entire surface specimen is etched away.

3.2.9 Field desorption or evaporation images

In addition to field ion images, some other types of images may be obtained in the field ion microscope or with the single atom detector.

It is possible to record field ion images well above the best image voltage or while the specimen is field evaporating. These types of images are generally used to increase the contrast between phases or to make planar features such as grain boundaries more evident. An example of a boron-decorated low angle boundary in a Ni_3Al specimen is shown in Fig. 3.28. The boundary is clearly visible in the micrograph taken during field evaporation but is not readily distinguishable in the field ion micrograph taken at the best image voltage.

Instead of forming the image from the ionized image gas, the image may be formed from the atoms in an adsorbed layer or from the substrate atoms. These types of images are known as field desorption and field evaporation images, respectively. However, these terms are often used interchangeably in the literature. An example of a field evaporation image is shown in Fig. 3.29. Field desorption and evaporation images are generally recorded in the absence of image gas and on a single atom sensitive detector such as in an imaging atom probe. Field desorption and evaporation images may be produced either by pulsed field evaporation or by slowly ramping the standing voltage on the specimen. These types of images may also be reconstructed from the three-dimensional atom probe data.

If field evaporation was perfectly uniform over the specimen surface, these images would show uniform intensity. However, several dark and bright lines along the zone axes are often evident in images of pure elements. The presence of these lines indicates that there are some local trajectory aberrations in the path of the ions. In addition, the central regions of poles are usually dark, indicating that the ion trajectories are deflected from the center of the

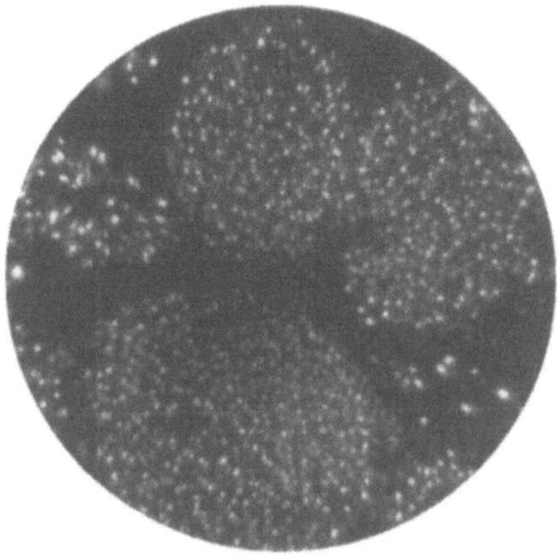

Fig. 3.27. Typical fracture surface of a field ion specimen of a nickel-based superalloy showing multiple emitting regions.

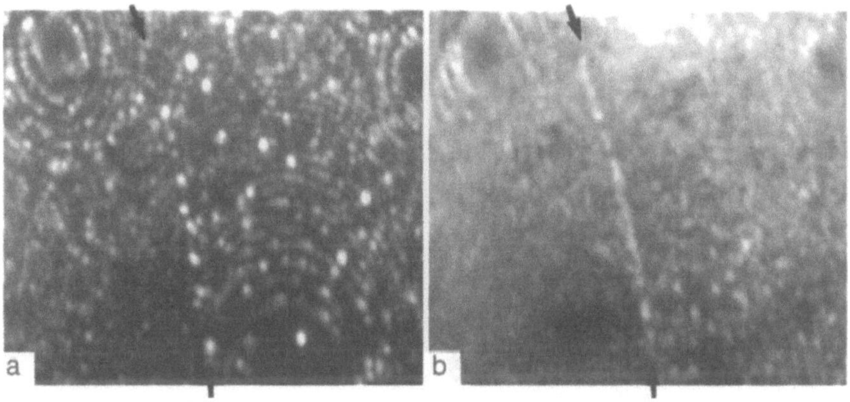

Fig. 3.28. Field ion micrographs boron-decorated low angle boundary in a boron-doped Ni_3Al specimen taken at a) the best image voltage and b) during field evaporation. The boundary is clearly evident in (b). From a collaboration with J. A. Horton, Oak Ridge National Laboratory.

pole. If the field evaporation images are recorded during the removal of many planes of atoms, persistent concentric rings around the poles are evident.

In materials that contain small precipitates or second phases, either bright or dark regions are produced in the field evaporation images. The contrast from precipitates is reversed from that in the field ion micrographs, that is precipitates appearing as dark regions in the field ion micrograph will appear as bright regions in the field evaporation images. The origins of this contrast

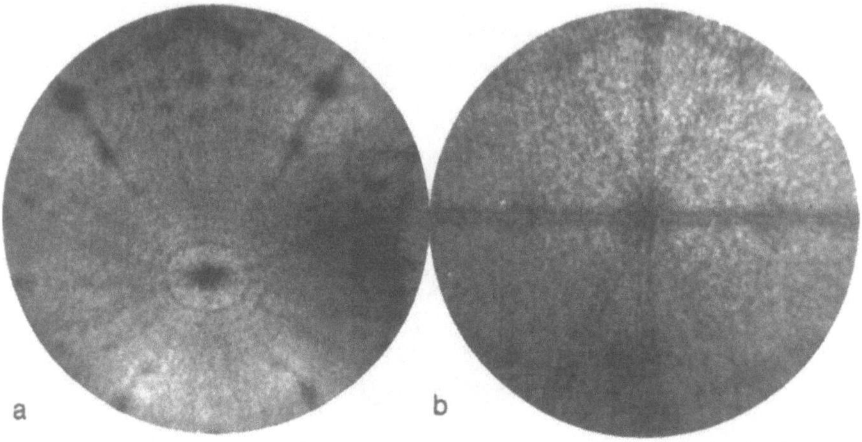

Fig. 3.29. Field desorption or evaporation image of a) tungsten and b) aluminum specimens showing bright and dark lines and dark regions at the center of the crystallographic poles.

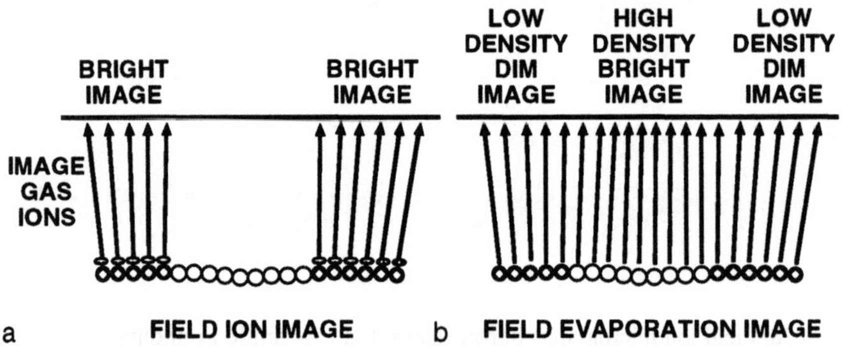

Fig. 3.30. Contrast mechanisms for field ion and field desorption images. a) in the field ion image, the ionization occurs preferentially above the phase that protrudes slightly from the mean surface of the specimen. b) in the field desorption image, the ion trajectories from these protruding regions are more divergent and result in a lower density of spots on the detector.

reversal are shown schematically in Fig. 3.30. As discussed in §3.2.3, a brightly-imaging precipitate will protrude slightly from the mean surface of the specimen. Therefore, the local radius of curvature of the precipitate is slightly lower than the surrounding matrix and the magnification is slightly higher.

Hence, the trajectories of the ions from the precipitate diverge more than the trajectories of ions from the matrix.

Field evaporation images may also be used to investigate the extent of trajectory aberrations at other microstructural features. For example, bright and dark lines are often observed along grain or other types of planar boundaries. The locations of the bright and dark regions have been observed to change after material was field evaporated from the specimen. This effect would appear to suggest that the local arrangement of the atoms at the boundary is influencing the trajectories of the ions.

3.2.10 Computer simulation of field ion and field desorption images

Computer simulation of field ion images is a useful method especially for materials with complex crystal structures. The most common application is the comparison of the field ion image with the simulations of different crystal structures for phase identification. Image simulations have been used with limited success in the visualization of point defects, dislocations, solid solutions and multi-phase materials [6]. One of the main limitations in simulations of these types of microstructural features is incorporation of surface relaxation.

Two types of simulations are the thin shell model and the bond model. In the thin shell model [7], the atoms that lie in a thin spherical shell on the surface of a specimen with a spherical end form are projected onto a flat surface with the use of one of the standard projections (i.e., stereographic, gnomic or orthographic). An example of computer simulation of a field ion micrograph of a body centered cubic crystal with a stereographic projection is shown in Fig. 3.31. Parameters that can be modeled include crystal structure (positions and type of atoms in the unit cell), specimen radius, thickness of shell, and orientation of the crystal.

The less frequently used bond model is based on the geometrical environment of the surface atoms [8]. The coordinates of all the atoms in the lattice are determined and only the atoms that are within a given radius are considered. The environments of each of these atoms are determined in terms of the number of first to sixth nearest neighbors. An imaging criterion based on the numbers of nearest neighbors is then imposed to produce the best match between the simulation and the experimental image. Only the atoms that satisfy this criterion are then projected onto a flat surface. This model has traditionally been applied only to small regions of the surface.

Field desorption and evaporation images have also been simulated and are discussed in §5.5.3.

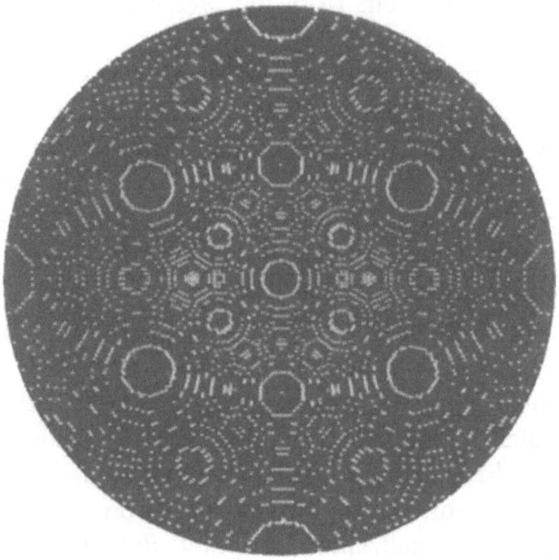

Fig. 3.31. Computer simulation of a field ion image generated with the thin shell model for a body centered cubic crystal with an 011 central pole.

3.3 Estimation of Parameters from Field Ion Images

Several important parameters may be measured directly from field ion micrographs. The specimen radius and the local taper angle of the specimen are important parameters in the reconstruction of three-dimensional data. The determination of these and other parameters that can be measured from individual or sequences of field ion images are discussed in this section.

3.3.1 Specimen radius

In materials that exhibit high quality field ion images, the radius of the specimen may be determined directly from the field ion micrograph [9]. This radius determination is performed by counting the number of atomic terraces or rings between the centers of two poles that are separated by a known angle. The method is shown schematically in Fig. 3.32 for a body centered cubic crystal. To obtain the maximum accuracy, the specimen should be field evaporated such that the last atom of the previous topmost plane should just have been removed and the new topmost plane is at its maximum diameter. The local radius of the specimen, r_t, is given by

$$r_t = \frac{nd_{hkl}}{1-\cos\theta} \qquad 3.3$$

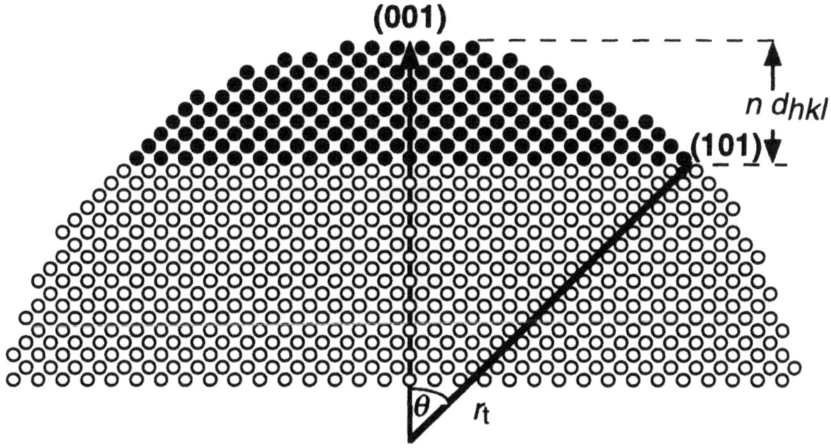

Fig. 3.32. Method to determine the specimen radius from the number of atomic terraces between known poles in a field ion micrograph.

where θ is the angle between the two planes, n is the number of counting planes between the poles, and d_{hkl} is the interplanar spacing of the counting plane. In the example shown, there are 10 planes between the (001) and the (101) planes. Therefore, $r_t = 10 \times 0.144 /(1 - \cos 45°) = 4.9$ nm. This method assumes that the interplanar spacing of the counting plane is accurately known. The formulae for calculating the interplanar spacing of a given *hkl* plane fromthe unit cell parameters (Appendix C) for a variety of common crystal structures is given in Appendix A. Caution should be exercised in ordered materials due to the contrast of different types of atomic planes that may obscure some of the planes. In materials that do not exhibit sufficiently high quality field ion micrographs, the radius of the specimen may be measured in a transmission electron microscope.

3.3.2 Taper angle

The average taper angle of the specimen may also be determined from a pair of field ion micrographs that were recorded after a known number of layers were field evaporated [5]. A schematic diagram of the method is shown in Fig. 3.33. The local taper angle of the specimen, $\alpha/2$ is given by

$$\sin \alpha/2 = \frac{r_2 - r_1}{r_2 - r_1 + n d_{hkl}} \qquad 3.4$$

where r_1 and r_2 are the radii of curvature of the specimen in the initial and final field ion micrographs, n is the number of planes field evaporated, and d_{hkl} is the interplanar spacing of the counting plane. This method may also be applied

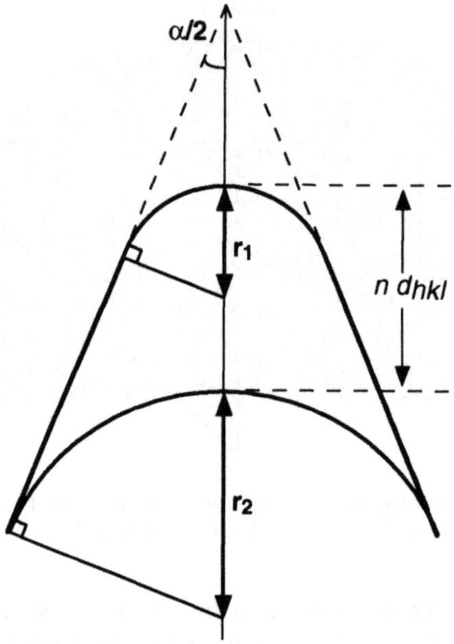

Fig. 3.33. Method to determine the taper angle from a field evaporation sequence by counting the number of atomic terraces removed.

when the two radii are determined from examination in the transmission electron microscope and when the number of planes or distance field evaporated between the images can be estimated directly from the three-dimensional data.

3.3.3 Image compression factor

The image compression factor may be defined as the ratio of the angle between two known crystallographic directions or poles in the specimen, θ, and the angle subtended between the same poles in the projected image on the field ion detector or the single atom detector, θ', as shown in Fig. 3.34. Therefore, the image compression factor, ξ, may be estimated from

$$\xi = \frac{\theta}{\theta'} \quad \text{and} \quad \tan\theta' = \frac{L}{\delta + r_t} \qquad 3.5$$

where L is the distance between the poles in the field ion image, δ is the distance from the specimen to the detector and r_t is the radius of curvature of the specimen. As the specimen radius is significantly less than the specimen to detector distance, r_t is normally neglected. A table of angles between poles for a cubic lattice is given in Appendix B.

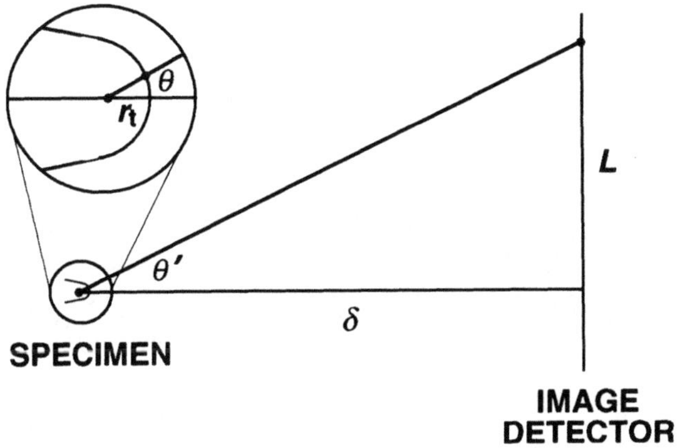

Fig. 3.34. Schematic diagram of the image compression factor. The distance of a pole from a second pole at the center of the image, L, the specimen to detector distance, δ, enables the angle θ' to be calculated and compared to the angle between the same poles in the crystal, θ.

3.3.4 Precipitate size - Persistence size method

The size of a small precipitate may be estimated from a field evaporation sequence by counting the number of planes removed while the precipitate is visible in the field ion image, as shown schematically in Fig. 3.35. This method estimates the extent of the precipitate along the specimen axis. If the precipitate is spherical, this method provides an estimate of the diameter of the precipitate. This persistence size method assumes that the interatomic spacing of the planes used to monitor the evaporation process is known. This procedure is usually performed from videotaped field evaporation sequences or from sequences of field ion images where each image is recorded after the evaporation of one plane so that the precise initial and final appearance of the precipitate can be accurately determined.

3.3.5 Angle between a planar feature and the specimen axis

The angle between a planar feature, such as a grain boundary, in a field ion micrograph and the specimen axis may be estimated from the micrograph [5], as shown in Fig. 3.36. The trace of the boundary is transferred to a stereographic projection such that the center of the field ion image is at the center of the projection and the scale and field of the view of the micrograph is matched to the diameter of the projection. The field of view of the field ion image is typically 100° compared to the 180° of the projection. The small circle through the boundary is completed. Extending the small circle outside the projection may be required to complete the circle. A line is constructed from the

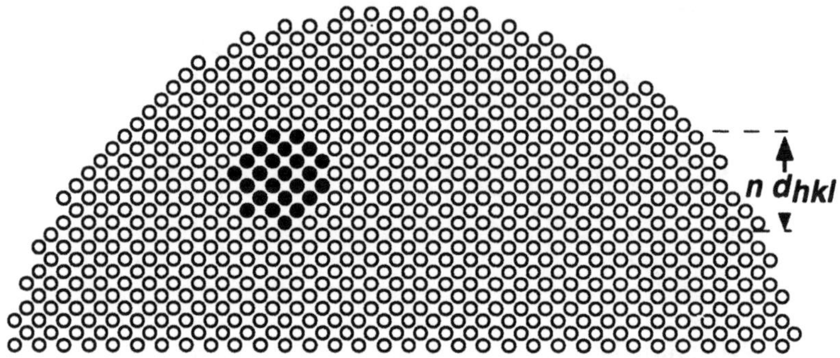

Fig. 3.35. Schematic representation of the persistence size method for estimating the size of a small precipitate. The number of planes field evaporated are counted from first appearance to final appearance of the precipitate in the field ion image.

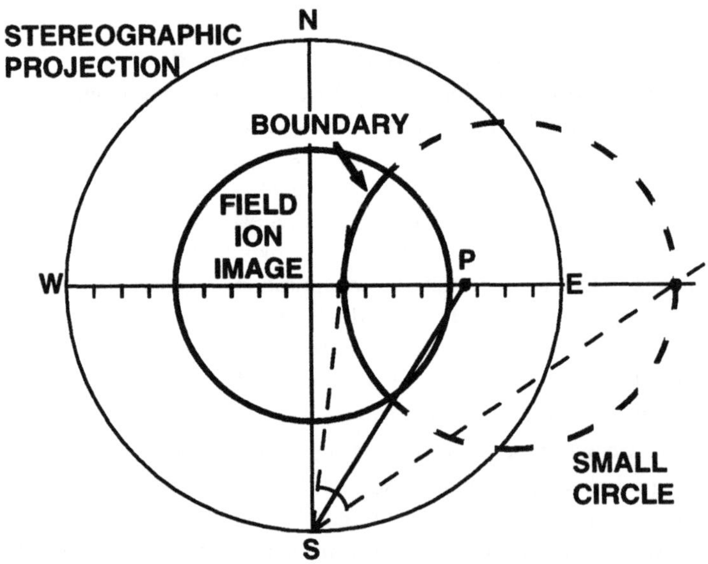

Fig. 3.36. Method to estimate the angle between the normal to the plane of a boundary and the specimen axis. See text for details.

center of the projection through the center of the small circle. This line defines the East-West axis of the projection. Two lines are constructed from the South pole of the projection, S, to the two intersections of the small circle with the East-West axis of the projection. The angle between these two lines is bisected at the South pole and the line SP is constructed to intersect the East-West axis. The point at which these lines intersect, P, is the normal to the defect plane and can be measured from the stereographic projection. The results of this method may be used to evaluate the accuracy of the reconstruction of three-dimensional atom probe data

3.4 Theoretical Background of Field Ion Microscopy

In this section, a brief outline of the theoretical background of field ion microscopy is presented. The theory of field ionization and field evaporation have been reviewed previously [1,10-20].

3.4.1 Theory of field ionization

The potential diagram for field ionization close to a clean metal surface is shown in Fig. 3.37. The potential seen by a tunneling electron at a distance x from the metal surface is given in the one-dimensional form by

$$V(x) = \frac{e^2}{4\pi\varepsilon_0 |x_i - x|} + eFx - \frac{e^2}{16\pi\varepsilon_0 x} + \frac{e^2}{4\pi\varepsilon_0 (x_i - x)}, \qquad 3.6$$

where F is the electric field strength, e is the elemental charge, ε_0 is the permittivity of free space, and x_i is the distance from the center of the positive ion to the electrical plane of the surface. In this equation,. the first term represents the potential well of the specimen. The second term is the external field. The third term is the image potential of the electron. The final term describes the repulsion of the electron by the image of the ion in the metal. As a gas atom approaches the metal surface, the rate of tunneling increases until a critical distance, x_c, is reached. At this point, the energy level of the electron in the gas atom coincides with the Fermi level in the metal. Tunneling is forbidden at low temperatures by the Pauli exclusion principle at distances closer than x_c, because no vacant states of appropriate energy exist within the metal to receive the electron. This region is known as the forbidden zone. The critical distance is given by

$$eFx_c = I - \phi - \frac{e^2}{16\pi\varepsilon_0 x_c} + \frac{\frac{1}{2}(\alpha_a - \alpha_i)F^2}{4\pi\varepsilon_0}, \qquad 3.7$$

where I is the ionization energy, ϕ is the work function of the surface, and α_a and α_i are the polarizabilities of the gas atom and ion (given in SI units of J $V^{-2}m^2$)†, respectively. The first two terms of this equation are evident from the form of Fig. 3.37. The third term is the image term and the final term is the polarization energy term. For most gases under typical field ion imaging conditions, the critical distance is approximately 0.45 nm. Therefore, field

† The Gaussian polarizability, measured in Å^3, is often used in the literature = $\alpha/4\pi\varepsilon_0$

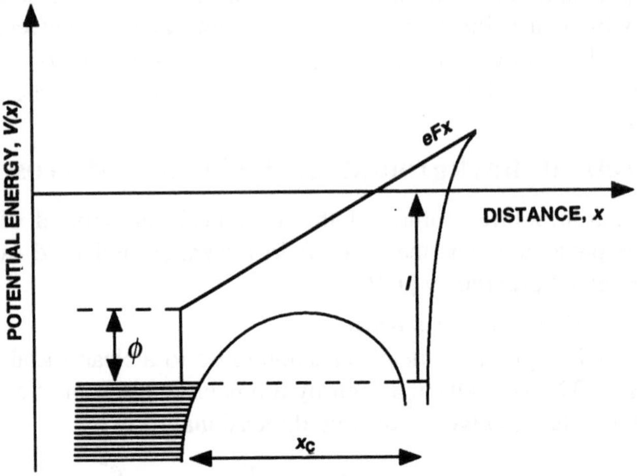

Fig. 3.37. Potential energy diagram for field ionization close to a clean metal surface.

absorbed layers of small atoms or molecules are not ionized under these conditions and remain bound to the surface of the specimen. In some cases, the absorbed layer may be removed by increasing the field. For example at 80K, absorbed layers of neon become unstable at ~48 V nm^{-1} and layers of molecular hydrogen are removed at ~33 V nm^{-1}. However, adsorbed layers of helium are stable up to the evaporation field of tungsten (57 V nm^{-1}).

The barrier penetration probability for field ionization may be expressed in terms of the Jefferies-Wentsel-Kramer-Brillouin (JWKB) approximation as [12,13]

$$D(E,V(x)) = \exp\left\{-\sqrt{\frac{32\pi^2 m_e}{h^2}}\int_{x_1}^{x_2}\sqrt{V(x)-E}\,dx\right\}, \tag{3.8}$$

where m_e is the mass of the electron, h is Planck's constant, $V(x)$ is the potential of the electron, E is the kinetic energy of the electron, and the integral extends over the barrier width. A simplified analytical solution of eqn. 3.8 was obtained by Gomer [14] under optimum tunneling conditions (i.e., when x_i=x_c) and by ignoring the image terms

$$D(x_c) = \exp\left\{-6.83\frac{I^{3/2}}{F}\sqrt{1-2.4\frac{\sqrt{F}}{I}}\right\}, \tag{3.9}$$

where I is expressed in electron volts and F in V nm^{-1}. This analytical solution and other numerical solutions have been used to predict the main features of field ionization, namely

1. There is a strong field dependence of the ionization probability. For example in the case of helium, the ionization probability increases by a factor of 10 when the field is doubled from 20 to 40 V nm^{-1}.
2. The total energy of the gas atom is always slightly lower than the applied voltage between the specimen and the ground plane. This indicates that the ions originate from a region a few tenths of nanometers above the surface of the specimen.
3. The energy distribution of the ion energy is extremely narrow at the best image field indicating that the ions are formed in an extremely narrow zone that is 0.015 to 0.025 nm in extent.
4. The overall ionization probability is substantially less than unity at normal imaging fields.
5. The ionization field for different gases depends on the 3/2 power of their ionization potentials.
6. There is an oscillatory structure in the tail of the field ion energy distribution, due to resonant tunneling effects.

3.4.2 Theory of field evaporation

In surface chemistry terminology, field evaporation is the removal of the substrate material and field desorption is the removal of adsorbed species from the surface.

Two basic theories of field evaporation have evolved. These theories are known as the image force model [12,13] (also known as the image hump or Müller-Schottky model) and the charge exchange model [14-19] (also known as the intersection or Gomer model). Both theories are based on simple one-dimensional models of atomic and ionic forces, image potentials and polarization effects. The ionic states of surface atoms are generally metastable with respect to the neutral state in the absence of an electric field. However, in the presence of an electric field, the ionic states become more stable as the distance from the specimen surface increases. At some point, the ionic and neutral potential energy curves cross and the removal of an ion by thermal activation over a reduced energy barrier is possible. The two models differ according to the sign of the gradient of the ionic potential energy curve at the point where it crosses the neutral state curve. The schematic potential energy diagrams for these two models are shown in Fig. 3.38.

In the image force model, the activation barrier for the desorption of an n-fold charged ion, under zero field conditions, $Q_{0(n)}$ is given by

$$Q_{0(n)} = \Lambda + \sum_{n} I_n - n\phi \,, \qquad 3.10$$

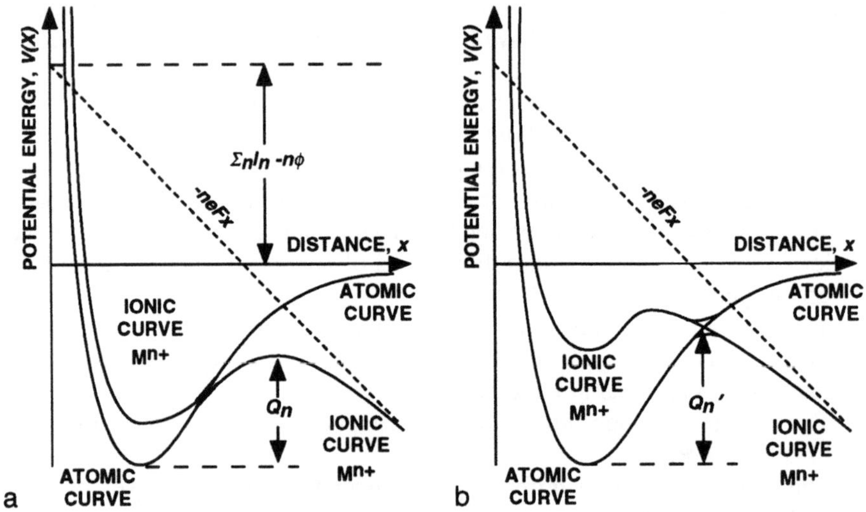

Fig. 3.38. Potential energy diagrams for a) the image force model and b) the charge exchange model. In the image hump model, the Schottky hump lies outside the atomic potential curve. In the charge exchange model, the Schottky hump lies inside the atomic potential curve.

where Λ is the heat of sublimation of a neutral atom, I_n is the nth ionization energy of the atoms and ϕ is the work function of the emitting surface. When an external electrical field is applied, the ionic states become more stable with increasing distance from the surface of the specimen and it becomes possible to desorb an ion over the reduced potential energy barrier. The position of the maximum on the ionic potential curve is given by

$$x_m = \frac{1}{2}\sqrt{\frac{ne}{4\pi\varepsilon_0 F}}\ , \qquad 3.11$$

where F is the field strength, ε_0 is the permittivity of free space and e is the elemental charge. The potential at this distance, which is known as the Schottky hump potential, is given by

$$-V_m = \sqrt{\frac{n^3 e^3 F}{4\pi\varepsilon_0}}\ . \qquad 3.12$$

Therefore, the energy barrier to desorption in the presence of an external electrical field is reduced to a value of Q_n, which is given to a first approximation by

$$Q_n = Q_{0(n)} - \sqrt{\frac{n^3e^3F}{4\pi\varepsilon_0}} \quad . \tag{3.13}$$

A correction may be applied to this equation to account for the change in the surface atom binding energy in the presence of the applied electrical field

$$Q_n = Q_{0(n)} - \sqrt{\frac{n^3e^3F}{4\pi\varepsilon_0}} + \frac{1}{2}c_aF^2, \tag{3.14}$$

where c_a is an empirical polarization energy coefficient for the atom. The rate constant K_n for field evaporation of an n-fold charged ion by thermal activation over the Schottky hump is expressed as an Arrhenius equation

$$K_e = \nu \exp\left(\frac{-Q_n}{k_BT}\right), \tag{3.15}$$

where ν is the vibration frequency of the surface atom. Similarly the time, τ, required for field evaporation is given by

$$\tau = \tau_0 \exp\left(\frac{Q_n}{k_BT}\right), \tag{3.16}$$

where τ_0 is the vibration time of the surface atom. From the above equations, the field required to evaporate an n-fold charged ion, F_n, is

$$F_n = \frac{4\pi\varepsilon_0}{n^3e^3}\left(Q_{0(n)} - k_BT\ln\frac{\tau}{\tau_0}\right)^2 . \tag{3.17}$$

From this image hump model, the evaporation field at absolute zero of any element may be estimated for different possible charge states. The most likely charge state exhibits the lowest evaporation field. The results of this type of calculation [17] are presented for most elements in Appendix C. The results from this model show reasonable agreement with experimental observations. As the evaporation rate is increased or the temperature of the specimen is decreased, the number of ions with higher charge states increases.

A more realistic approach may be taken for the field desorption of neutral absorbates and for low temperature field evaporation of metals with the charge exchange model. In this model, the position of the Schottky hump lies inside the potential energy curve of the neutral state. If the field is sufficient to remove the hump, the surface atom may be field evaporated by an energy level corresponding to the intersection of the neutral and ionic curves. The activation energy, Q_n', is lower than that predicted from the size of the Schottky hump and is given by [15]

$$Q_n' = Q_{0(n)} - \frac{n^2e^2}{16\pi\varepsilon_0 x_{\text{int}}} - neFx_{\text{int}} - \frac{B_a}{2} - \Delta E_a + \mu_a F + \frac{\frac{1}{2}(\alpha_a - \alpha_i)F^2}{4\pi\varepsilon_0}, \quad 3.18$$

where x_{int} is the distance from the electrical surface of the metal at which the potential curves intersect, B_a and ΔE_a represent the broadening and shift of the adsorbate energy level due to the interaction with the substrate, μ_a is the zero field dipole moment of the adsorbate, and α_a and α_i are the polarizabilities in the neutrally bound and ionic states, respectively. This model has been found to yield approximately the same field evaporation fields at absolute zero as the image hump model [18].

3.4.3 Post ionization

It is possible for field evaporated ions to undergo further field ionization to higher charge states after leaving the surface because the potential energy curves for higher ionic charge states cross the potential energy curves of the lower charge states at sufficiently large distances from the specimen [20-23]. This field ionization process is known as post-ionization and is generally regarded as the most important mechanism for the production of multiply charged ions. The predictions from this model are in good agreement with experimental observations for a large number of elements. As the post-ionization process occurs over an extended region, a broader energy distribution is experimentally observed for multiply charged ions.

References

1. M. K. Miller, A. Cerezo, M. G. Hetherington and G. D. W. Smith, Atom Probe Field Ion Microscopy, Oxford University Press, Oxford, UK, 1996, p. 134.
2. Y. C. Chen and D. N. Seidman, *Surf. Sci.*, **26** (1971) 61.
3. C. M. C. de Castilho and D. R. Kingham, *J. Phys. D Appl. Phys.*, **20** (1987) 116.
4. T. J. Wilkes, G. D. W. Smith and D. A. Smith, *Metallography*, **7** (1974) 403.
5. K. M. Bowkett and D. A. Smith, Field Ion Microscopy, North Holland, Amsterdam, The Netherlands, 1970.
6. A. J. Moore, in Field Ion Microscopy, eds. J. J. Hren and S. Ranganathan, Plenum Press, New York, NY, 1968, p. 69.
7. A. J. Moore, *J. Phys. Chem. Solids*, **23** (1962) 907.
8. R. C. Sanwald and J. J. Hren, *Surf. Sci.*, **7** (1967) 197.
9. M. Drechsler and P. Wolf, Proc. 4th Intl. Conf. Electron Microscopy, Berlin, **1**, Springer-Verlag, Berlin, 1960, 835.
10. R. G. Forbes, *Appl. Surf. Sci.*, **87** (1995) 1.
11. R. G. Forbes, *Appl. Surf. Sci.*, **94/95** (1996) 1.
12. E. W. Müller, *Phys. Rev.*, **102** (1956) 618.
13. E. W. Müller, *Adv. Electron. Electron Phys.*, **13** (1960) 83.

14. R. Gomer, *J. Chem. Phys.*, **13** (1959) 341.
15. R. Gomer and L. W. Swanson, *J. Chem. Phys.*, **38** (1963) 1613.
16. L. W. Swanson and R. Gomer, *J. Chem. Phys.*, **39** (1963) 2813.
17. T. T. Tsong, *Surf. Sci.*, **10** (1968) 102.
18. T. T. Tsong and E. W. Müller, *Phys. Status Solidi A.* **1** (1970) 513.
19. K. Chibane and R. G. Forbes, *Surf. Sci.*, **122** (1982) 191.
20. N. Ernst, *Surf. Sci.*, **87** (1979) 469.
21. N. Ernst and T. Jentsch, *Phys. Rev. B*, **24** (1981) 6234.
22. R. Haydock and D. R. Kingham, *Phys. Rev. Lett.*, **44** (1980) 1520.
23. D. R. Kingham, *Surf. Sci,*, **116** (1982) 273.

Chapter 4

Instrumentation

In this chapter, the various hardware components of a three-dimensional atom probe are described in detail. There is a wide variation in the precise construction of the different three-dimensional atom probes that have been designed. Therefore, some of the items discussed in this chapter may not be available on all instruments. It is important to understand the operation and limitations of these components in order to optimize the experimental parameters and to prevent damage to the equipment.

4.1 Vacuum System

The vacuum system of a modern atom probe is designed to operate at ultra high vacuums (UHV) in the 10^{-11} mbar (10^{-9} Pa) range. Therefore, the construction and all internal components used in its fabrication must be capable of operating under these conditions without significant outgassing. The vacuum vessel is invariably constructed from welded stainless steel tubes and Conflat® metal seal flanges. Typical materials used in the vacuum system are stainless steel, copper, nickel, alumina, sapphire and machineable glass. Small quantities of Kapton® or polytetrafluoroethylene (PTFE) insulation and Viton® seals in some types of valves may also be used. To achieve the desired base pressure, it is normal to bake the entire system to a temperature of up to ~200°C for several hours after the vacuum is either exposed to atmosphere or periodically during normal use. The interval between bakes is dependent on the usage and the degradation of the base pressure of the system. The maximum bakeout temperature is primarily limited by the microchannel plates, which lose gain if baked to higher temperatures. The valves and manipulators may impose additional maximum temperature limitations.

Most three-dimensional atom probes utilize a three chamber approach: the main analysis chamber that includes the specimen stage, cryostat, field ion imaging system and the mass spectrometer; an optional preparation chamber where the specimens are stored in a multi-specimen carousel prior to examination; and a fast entry airlock which is used to introduce and remove specimens from the system. Some additional items such as ion guns, environmental chambers, and thin film deposition equipment can be fitted to the preparation chamber. Specimens are transferred between the chambers with the use of standard UHV-compatible rack-and-pinion or magnetically coupled transfer devices.

The vacuum vessel is evacuated with either turbomolecular pumps or liquid nitrogen trapped oil diffusion pumps. These primary pumps are backed by rotary or diaphragm pumps. It is important that backstreaming of the oil in the rotary pumps is eliminated especially in systems with turbomolecular pumps as there is no liquid nitrogen trap. Therefore, foreline traps are incorporated between the rotary pumps and the main pumps. Most instruments are also equipped with titanium sublimation pumps. These sublimation pumps should be fired every few hours depending on the base pressure of the system in order to maintain low levels of active gases such as hydrogen. Some instruments are also equipped with auxiliary ion pumps. The cryostat can also provide additional cryopumping by condensing some gases on the cold surface of the cryostat. However, these gases desorb back into the system when the cryostat is heated.

The pressure in the main chamber is normally measured by either an ion gauge or an inverted magnetron gauge. The ion gauge has the disadvantage that the light and electrons produced during operating can interfere with the single atom detector and generally cannot be operated during the experiment. However at the present time, ion gauges are capable of reading lower base pressures than inverted magnetron gauges. The pressures in the other chambers are measured by a combination of ion, cold cathode, Penning, pirani or thermocouple gauges depending on the pressure range. In modern instruments, it is normal to interlock the readings from these gauges with the timing and imaging electronics to prevent operation of the detectors in the event of a pressure increase. The pressures read by these gauges can also be monitored during an analysis by the computer that controls the experimental operation of the atom probe.

4.2 Field Ion Microscope

In this section, the components that comprise the field ion microscope subsystem of the atom probe are described.

4.2.1 Specimen stage

The stage is capable of rotating the specimen about its apex so that all regions of the specimen are available for analysis in the mass spectrometer, as shown in Fig. 4.1. Typical ranges of rotation required for this capability are at least ±45° in both horizontal and vertical directions. Additional ranges of rotation allow the specimen to be aligned with the field ion imaging system, the mass spectrometer and the specimen exchange positions. In most atom probes, these two orthogonal rotations are accomplished with the use of a combination rotary and linear drive. In the tomographic atom probe, the specimen is rotated around one axis and the second rotation is about the specimen axis. This alternative method provides equivalent access to all regions of the specimen but it changes the algorithm for the reconstruction of the data when the specimen is rotated from the ion-optical axis of the atom probe. In the classical atom probe, the specimen also had to be translated to align the specimen with the optical axis of the mass spectrometer. These additional translations are not normally provided in three-dimensional atom probes but care is exercised during the loading of the specimen to ensure that the needle is straight, aligned with and set at the correct position along the central axis of the instrument. These rotations are only macroscopically about the center of the hemispherical cap of the specimen, since the position of the apex of the specimen cannot be controlled to better than approximately ± 0.25 mm. The stage normally incorporates a simple locking mechanism to prevent the specimen from drifting during analysis. Unlike most other high spatial resolution techniques, vibration damping or isolation is not required in a three-dimensional atom probe, as the imaging process does not magnify the magnitude of the vibrations.

The specimen stage is designed to accept interchangeable specimens. A common method of specimen preparation is to carefully crimp the field ion needle into an electropolished copper or nickel tube. Some brittle specimens are glued into place with a special UHV-compatible silver-containing conducting epoxy. Care should be taken not to create a trapped volume that will act as a virtual leak when placed into the ultrahigh vacuum system. The copper or nickel tube is then inserted into a threaded nickel stub and held in place with a set screw. This loading operation is usually performed under a low power (~30x) optical microscope to ensure that the specimen is set at the correct position. The importance of the correct specimen position is discussed in §5.4. Nickel is generally used for the interchangeable stub since it has good thermal and electrical conductance at cryogenic temperatures and it can be held in place in the airlock or carousel with a permanent magnet. In addition, the use of a nickel stub and a copper stage minimizes the problem of cold welding the stub into the stage. Ultrahigh vacuum compatible lubricants, such as a suspension of molybdenum disulphide powder in methanol or acetone, are also used sparingly on the treads of the specimen stub to prevent cold welding. Good

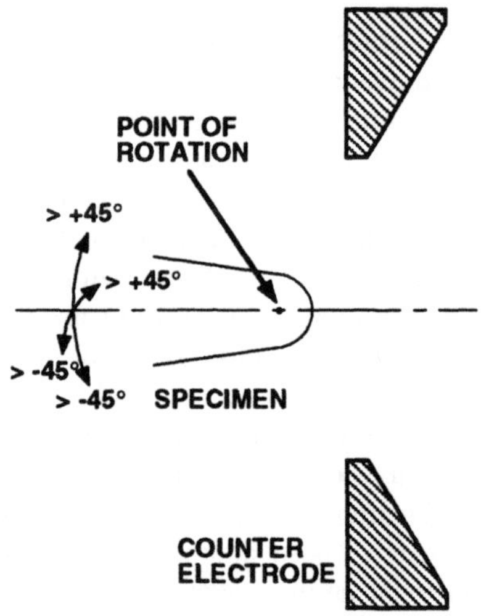

Fig. 4.1. The specimen may be rotated about its apex region by at least ±45° in both horizontal and vertical directions so that all regions on the surface may be aligned with the mass spectrometer.

metal-to-metal contact between all these components ensures that the specimen attains the desired cryogenic temperature.

The nickel stub holding the specimen is then loaded into the airlock and when evacuated to a suitable pressure is transferred from the airlock to the specimen storage carousel in the preparation chamber. Finally, the nickel stub can be moved from the carousel and screwed into the stage.

The stage is designed to isolate the specimen both electrically and thermally from the vacuum system so that a high voltage may be applied to the specimen (§4.2.3) and the specimen may be cooled to cryogenic temperatures. This insulation is generally achieved with alumina, machineable glass, sapphire or boron nitride spacers. Alumina or machineable glass are used where both electrical and thermal insulation are required. Sapphire or boron nitride are used where electrical insulation and good thermal conduction are required. The specimen region of the stage is generally attached to the cryostat with one or more flexible copper or gold braids.

4.2.2 Cryostat

The standard operating temperature of the specimen varies from below 15 K to approximately 70 K depending on the material. Selection of the correct operating temperature during atom probe analysis is discussed in §6.3.

The specimen temperature for field ion microscopy may be chosen to optimize the contrast between phases (§3.2.4).

The design of cryostats for atom probes has evolved from static reservoirs of cryogenic liquids such as liquid nitrogen, through hydrogen and nitrogen cryotips, to commercial closed-cycle helium cryogenerators. Closed-cycle cryogenerators are based on the gas expansion cycle. In this cycle, helium is first compressed and the heat of compression is removed by a heat exchanger. The compressed gas is subsequently allowed to expand and thereby extracts the heat from the thermal load attached to the cryostat.

A two stage cryogenerator is normally used in atom probes, as shown in Fig. 4.2. The first stage of the cryogenerator is typically operated at a base temperature of ~77K and is used to cool a copper shroud surrounding the specimen. This shroud has access holes for the specimen exchange system and the various detectors. This shroud minimizes the radiation heating of the specimen from the walls of the chambers and also provides a ground plane to minimize stray electrical fields from the specimen. The specimen is cooled by the second stage of the cryogenerator, which typically operates at a base temperature of 10 to 20 K depending on the cooling power of the cryogenerator. The thermal mass of the specimen stage and radiation shield are minimized so that the time taken to cool the specimen and the time required to change the temperature of the specimen are kept to a minimum. The cryostat is continuously operated at cryogenic temperatures in most instruments. Some of the external components of the cryogenerator, such as the motor and internal piston assembly, must be removed prior to baking the system.

The specimen is maintained at the selected temperature by passing a current through a heating element attached to the base of the cryostat. This heating element is normally fabricated out of an insulated high resistance wire or an UHV-compatible resistor. When the cryostat is heated, a significant increase in pressure may be generated due to desorption of gases that were condensed onto the cryostat. It is normal practice after extended operation at low temperatures to increase the temperature of the cryostat to desorb these condensed gases.

The temperature is measured with a sensor attached to the base of the cryostat. Therefore, the temperature of the specimen is often a few degrees warmer than that indicated by the sensor due to the different positions of the sensor and the specimen and the thermal conduction from the support structure of the stage.

The heating element and the temperature sensor are connected to a temperature controller. In some cases, the compressor of the cryogenerator is interfaced to the temperature controller so that the cycle rate of the compressor may be optimized. The temperature controller may also be connected to the control computer with a serial (RS-232 or similar) or an IEEE-488 interface.

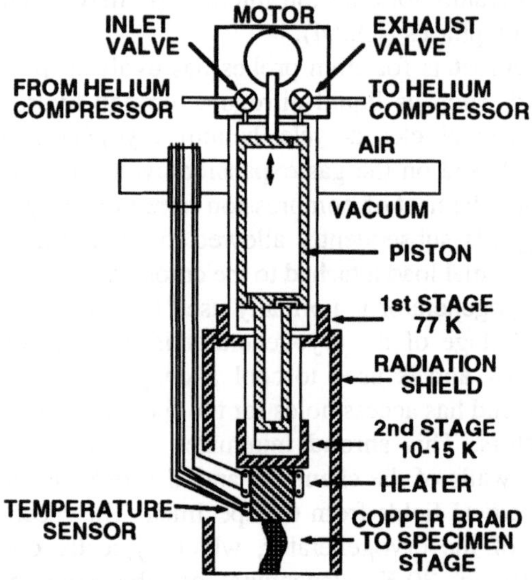

Fig. 4.2. Schematic diagram of a two-stage closed-cycle cryogenerator based on the gas expansion cycle. The first stage is used to cool a radiation shield around the specimen and the second stage the specimen. A heating element and temperature sensor are attached to the bottom of the second stage.

4.2.3 High voltage system

The high voltage system consists of a variable high voltage power supply for the standing voltage on the specimen, a counter electrode, a high voltage pulser, and a 50 Ω load, as shown in Fig. 4.3. The operating standing voltage is up to +30kV depending on the sharpness of the specimen and could in principle be extended to higher voltages for blunt specimens. However, most instruments limit the maximum voltage to 30 kV to avoid the generation of X-rays. A high impedance (~100 MΩ) resistor is normally incorporated into the high voltage feedthrough on the standing voltage to limit the current drawn from the supply during high voltage discharges which might damage the power supply. During normal operation, the power required for producing the field ion image is negligible. A negative voltage may also be applied to the specimen to produce field electron emission images. An additional variable 0 to +30 kV high voltage power supply is required for the energy-compensating reflectron lens (§4.3.1).

To field evaporate the ions from the surface at a well defined time, a high voltage pulse is applied to the specimen. This pulse is superimposed on the standing voltage on the specimen. Two configurations are in common usage,

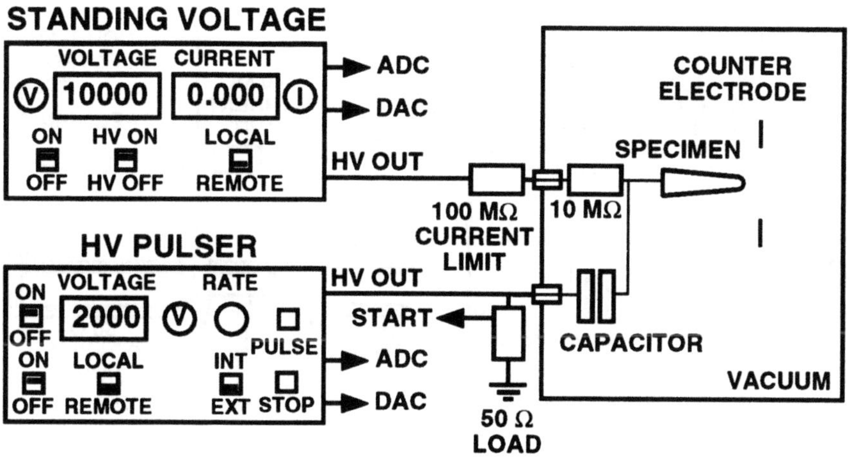

Fig. 4.3. Schematic diagram of the high voltage system. The connections to the specimen are designed so that they do not interfere with the movements of the specimen stage.

as shown in Fig. 4.4. In the first configuration, a positive pulse is capacitively coupled to the standing voltage. The connection is made as close to the specimen as possible with an internal UHV-compatible capacitor and usually incorporates a 50 Ω load on the vacuum feedthrough. The internal capacitor is generally fabricated from a short length of alumina tube that has portions of both internal and external surfaces coated with an electrically conducting film. A grounded circular counter electrode is positioned slightly in front of the specimen. This configuration usually includes a short length of impedance mismatched line to the specimen that degrades the pulse that reaches the specimen. In the second configuration, a negative pulse is applied to the counter electrode through an internal impedance-matched 50 Ω coaxial cable. A second coaxial connection from the counter electrode enables the pulse to be transmitted back to a 50 Ω load located on a vacuum feedthrough. This negative pulse configuration enables the proper matching of the impedance to be maintained throughout the pulse line and can improve the mass resolution. However, in this negative pulse configuration, some of the amplitude of the pulse is lost due to coupling of the pulse to the specimen. Typical pulse coupling factors are in excess of ~85%. In both cases, the 50 Ω load terminates the pulse line and also incorporates a 50 Ω source (or pickoff) that is used as the start signal for the timing system (§4.3.4).

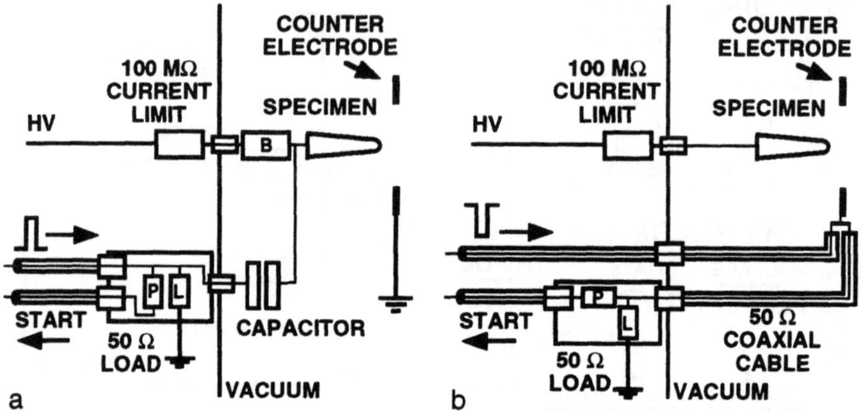

Fig. 4.4. High voltage configurations: a) positive pulse applied to the specimen and b) negative pulse applied to a counter electrode in front of the specimen.

Several methods have been adopted for motion of the counter electrode with respect to the specimen. In the optical atom probe, the specimen and the counter electrode rotate together and the relative positions do not change. This arrangement ensures the same field distribution on the apex of the specimen for all analysis positions. In the tomographic atom probe, the counter electrode is maintained at a fixed position and the specimen rotates. The electrical connections to the counter electrode are significantly simpler in this arrangement as the motions of the stage do not have to be taken into account. A hybrid arrangement is used in the energy-compensated optical position-sensitive atom probe where the counter electrode is fixed in the vertical plane but moves with the specimen in the horizontal direction.

It is good practice to cryogenically cool the counter electrode to approximately the same temperature as the specimen as it is usually the closest object to the specimen. If the counter electrode is at a significantly higher temperature, it may act as a heat source and increase the temperature at the apex of the specimen.

Prior to the introduction of commercially available solid state pulsers in the mid 1990's, the high voltage pulse was produced by charging a capacitor and then closing a mercury wetted switch to discharge the stored energy into a load [1], as shown in Fig. 4.5. These mechanical mercury-wetted reeds were limited to operating voltages of up to ~2.5 kV and a maximum operating frequency of between 50 and 200 Hz, depending on the size of the reed. In contrast, the solid state pulsers have output voltages of up to 4 kV and may be operated at frequencies of up to 2 kHz. These solid state pulsers are based on

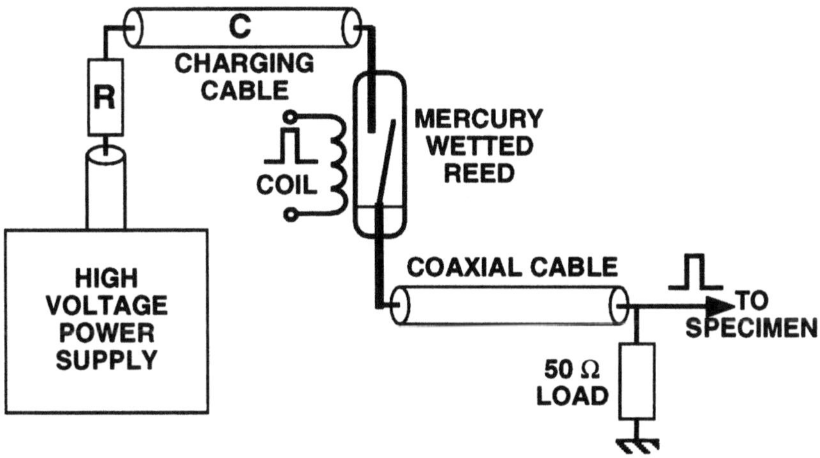

Fig. 4.5 A mercury wetted reed high voltage pulser provides the high voltage pulse required for field evaporation. The length of the pulse (10-20ns) is determined by the length (i.e., capacitance) of the charging cable.

transistor switches that are made up of a large number of metal-oxide semiconductor field effect transistors (MOSFET) arranged in parallel and in series. Each MOSFET is controlled synchronously via a gate with special circuits within the switch. A schematic diagram of a solid state pulser is shown in Fig. 4.6. The typical rise time of the pulse from a mercury wetted reed is slightly faster (< 1 ns) than that of the solid state variant (~1.2 ns). Therefore, instruments equipped with mercury wetted reeds should exhibit slightly better mass resolution. In practice, the significantly faster operating frequency of the solid-state pulser normally outweighs the minor loss of mass resolution. The operating frequency of these high voltage pulses is one of the main limiting factors in the data accumulation rate of the instrument. Factors affecting the performance of the pulser are the stability and the degree of jitter of the pulse amplitude and the repeatability of the pulse shape. In the mercury-wetted reed designs, an additional circuit is often included to prevent the generation of a second lower amplitude pulse, known as an afterpulse [2,3]. This afterpulse is generated because the capacitor is able to recharge a small amount in the time interval between the initial plasma discharge and the physical impact of the mechanical contacts.

An alternative system may also be used to remove atoms from the surface by replacing the high voltage pulse with a short duration laser pulse, as first described by Kellogg and Tsong in 1980 [4]. Care is required to shield

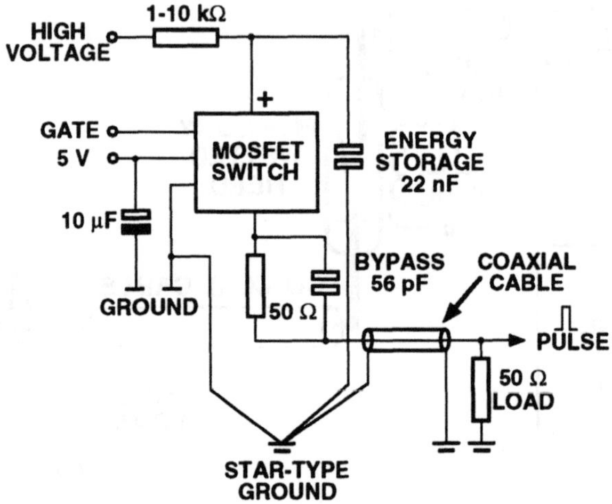

Fig. 4.6. Solid state pulser based on a MOSFET switch that can produce high voltage pulses of up to 4 kV at pulse repetition rates of up to 2000 Hz. Positive pulse configuration is shown. For negative pulse configuration, the polarity of the MOSFET switch is reversed.

reflections from the laser pulse from the single atom detector since it is sensitive to the photons. This alternative instrument known as a pulsed laser atom probe (PLAP) is generally applied only to specimens with poor electrical conductivity such as semiconductors and some conducting ceramics.

Due to the long duration of a typical atom probe experiment, automatic control of the voltages is necessary to compensate for the change in the radius of curvature of the specimen as material is field evaporated. Therefore, the high voltage power supplies and the high voltage pulser are all interfaced to the control system by analog-to-digital (ADC) and digital-to-analog (DAC) converters. This configuration enables the voltages to be read and set by the control computer (§4.3.5).

4.2.4 Image gas

Most instruments are equipped with several image gases so that a range of materials may be characterized. The common choices of image gases are neon, helium, argon and hydrogen. The main parameters of the image gases are summarized in Table 4.1. The temperature of the specimen and the cryostat must be taken into consideration during field ion microscopy so as not to condense the image gas on the cold surfaces. Neon is used for most transition elements and alloys and helium is used for the more refractory elements. Argon and hydrogen are sometimes used at the start of an experiment to evaporate through oxide or other surface films due to their low ionization fields. However, argon and hydrogen are normally avoided, as they are difficult to remove from the vacuum system. Hydrogen also combines with metal atoms to form multiple hydride species that complicate the interpretation of the mass spectra (§5.4). The presence of residual hydrogen in the vacuum system may be indicated by a low intensity hydrogen field ion image appearing at approximately 60% of the field required to form a neon image. Several other image gases such as nitrogen, oxygen and carbon monoxide have also been used for special applications. In rare cases, image gas mixtures may be used to improve the quality of the field ion images.

The presence of the image gas also reduces the field required to field evaporate the specimen. Therefore at the start of an atom probe analysis, it is common to progressively increase the voltage on the specimen after the image gas is removed from the system to initiate field evaporation. Conversely, if a specimen is imaged after analysis in the atom probe, the standing voltage may have to be reduced to prevent excessive field evaporation in the presence of the image gas.

A typical configuration for the gas handling system is shown in Fig. 4.7. The image gas is usually supplied in the form of 1 or 2 liter glass flasks that have to be glass blown onto a glass-to-metal feedthrough. These flasks are supplied with a thin glass seal that is broken by a steel ball bearing after the dead space has been evacuated, as shown in Fig. 4.7. Alternatively, a metal gas cylinder is attached to the system by means of a gas regulator. The purity of the image gas is typically research grade (99.995% or better). In the case of gas cylinders, a filter may be incorporated into the gas lines to remove impurities from the image gas. Generally, each image gas is admitted into a manifold at a lower pressure than the primary container so as to reduce the pressure differential across the leak valve to the analysis chamber. The manifold and its connections to the pumping system also permit the evacuation of the dead space between the glass flask and the manifold after a new flask is attached to the system. The typical image gas pressure used for field ion microscopy is 1 to 5 x 10^{-5} mbar. This pressure is defined by the brightness of the field ion image, higher pressures producing brighter images. If the gas pressure is too high, some artifacts such as comets (§3.9) will form. Excessive pressures can also result in severe damage to microchannel plates and phosphor screens from electrical discharges. Single atom detectors should not be operated at normal gain levels in the presence of the image gas. However, field ion images may be formed on the single atom detector when operated at significantly reduced gain. The sensitivity of an ion gauge is different for each image gas and should be corrected by the gauge factor, as shown in Table 4.1. Some modern ion gauge controllers have this correction incorporated into their design.

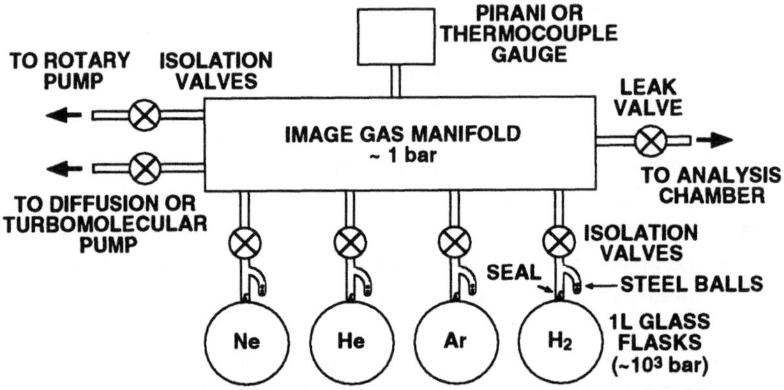

Fig. 4.7. The image gas handling system permits a choice of image gas (neon, helium, argon or hydrogen) to be admitted to the analysis chamber through a leak valve. The system also permits the manifold and the dead volumes to be evacuated so that the glass flasks may be exchanged when exhausted.

Table 4.1. Selected parameters of image gases.

Image Gas	Image Field V nm^{-1}	Boiling Point K	Ion Gauge Factor
Hydrogen	22.8	20.3	0.42-0.53
Helium	44.0	4.5	0.25
Neon	37.0	27.0	0.29
Argon	23.0	87.2	1.4
Nitrogen	17	77.4	1

The image gas may be admitted to either a static vacuum (i.e., where the pumps are isolated from the main chamber) or to a dynamically-pumped system (usually with some reduction of the pumping speed of the pumps). In the static vacuum, an increase in the background pressure during the imaging process may lead to artifacts if the experiment is conducted over a long period of time in a system with a significant leak rate. In the dynamically pumped system, some impurities are constantly introduced into the system with the image gas and significantly larger quantities of image gas are consumed and therefore, larger capacity gas cylinders are used.

4.2.5 Imaging system

The field ion imaging system consists of an image-quality microchannel plate (MCP) image intensifier and a phosphor screen, as shown in Fig. 4.8.

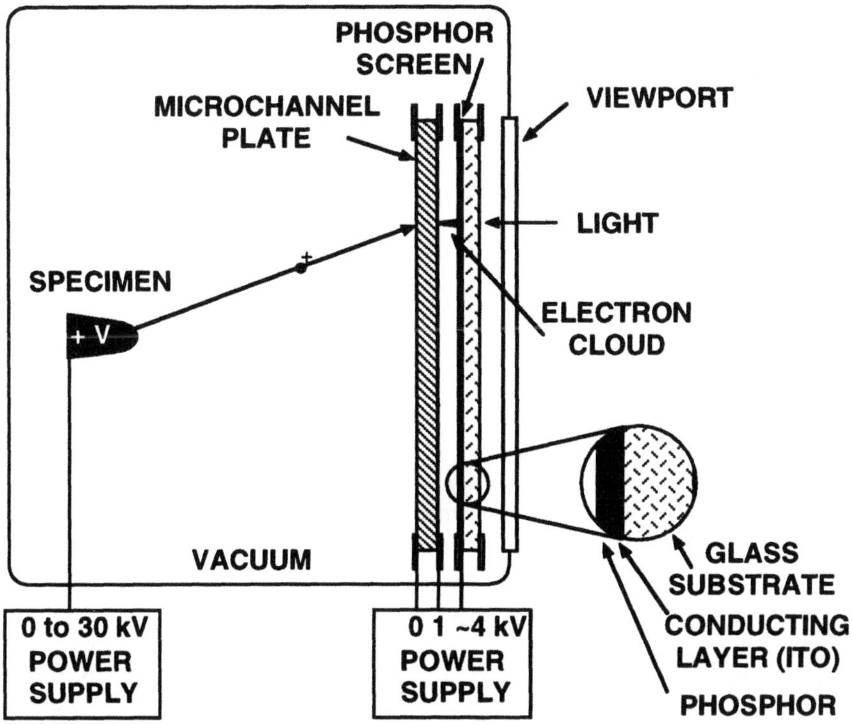

Fig. 4.8. The field ion imaging system consists of a single microchannel plate positioned ~1-2 mm in front of a phosphor screen. The light output from the phospor is viewed through an optical quality viewport on the vacuum system.

The use of microchannel plates reduced lengthy exposure times originally required to record field ion images directly on phosphor screens to the order of a second or less without degrading the spatial resolution. A microchannel plate is a thin disk that consists of a two-dimensional close packed array, similar to a honeycomb, of 8 to 25 μm diameter glass tubes whose internal surfaces have a high secondary electron yield, as shown in Fig. 4.9. Microchannel plates are extremely delicate and therefore require careful handling especially during installation and first time operation. In addition, they should be stored under a moisture- and oil-free high vacuum when not in use. The finer diameter tubes provide higher spatial resolution and are normally used in imaging applications. The front and rear surfaces of the disk are coated with a thin layer of conducting material, such as nichrome, so that when a voltage is applied across the surfaces there is an increasing positive field along each channel.

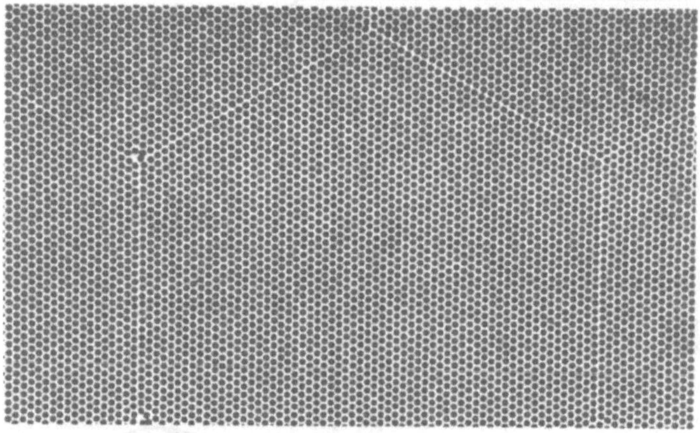

Fig. 4.9. Optical micrograph of the channels in a microchannel plate. The channels are arranged in a close packed arrangement. The coarse hexagonal pattern is the outline of one of the channel bundles fabricated as an intermediate step in the manufacturing process.

When an ion enters one of the glass tubes, it strikes the internal surface and produces a few electrons. These electrons are then drawn along the tube due to the increasing field and produce more electrons each time they strike the surface thus amplifying the signal. The magnitude of this electron multiplication process or gain of the microchannel plate is a function of the voltage applied between the front and rear surfaces. The voltage applied (typically ~0.8 to ~1.2 kV) across a single microchannel plate is normally adjusted to produce ~10^3 electrons for each incident ion. Since the distance between the microchannel plate and the phosphor screen is typically 1-2 mm and a bias voltage (typically ~3 kV) is applied between the microchannel plate and the phosphor screen to minimize any significant lateral spread of the electron cloud, the field ion image is proximity focused onto the phosphor screen.

The phosphor (P20 or similar) is deposited onto a glass plate that is coated with a transparent electrically conducting film such as tin oxide or indium tin oxide (ITO). In addition to increasing the gain, microchannel plates also convert the incident ions into electrons. The ion-to-electron conversion in the microchannel plate has several advantages as electrons are significantly more efficient in producing light (~50% compared to ~1 % for helium) and do not damage the phosphor. Significant damage can occur to the imaging system if it is operated at high pressure (≥10^{-4} mbar).

The field ion image can be recorded on either 35 mm film (e.g. 400 E.I./ISO), a digital camera (sensitivity ~0.25 lux at f1.2), or by capturing the signal from a high resolution video camera with a frame grabber in a computer. In all cases, cameras that permit interchangeable wide aperture (f1.2-1.4) lenses and extension tubes are generally required in order to fill the frame with the full field ion image since the distance from the film plane to the phosphor screen is restricted by the vacuum system.

4.3 Mass Spectrometer

The time-of-flight mass spectrometer is a field free drift tube between the specimen where the ions are emitted and the primary detector where they arc collected. A time-of-flight type of mass spectrometer is used as it has an unlimited mass range and does not have to be tuned to a specific mass window. The mass spectrometer may also include an energy-compensating lens to improve the mass resolution. Care has to be taken in the placement of items such the high voltage connections to the specimen and ion or inverted magnetron gauges, which contain voltage sources, so as not to interfere with the field free region.

4.3.1 Energy compensation

The main source of the limitation in mass resolving power (often referred to as mass resolution) is the small energy deficit that is associated with the high voltage pulse that is used to field evaporate ions from the specimen. This energy deficit arises since ions may field evaporate at different times during the application of the high voltage pulse and therefore acquire slightly different energies before they reach the field free region. Although energy compensation schemes were introduced to the classic atom probe by Müller and Krishnaswamy and as early as 1974 [5] and were in common use in atom probes in the early eighties, they were not incorporated into the first three-dimensional atom probes.

The initial method of energy compensation used in atom probes was the addition of an electrostatic lens between the specimen and the single atom detector. This electrostatic lens was a sector of a toroid and was based on a design described by Poschenrieder [6]. However, this configuration was not appropriate for the large acceptance angles required in a three-dimensional atom probe. Subsequently, a reflectron lens or energy mirror was used [7-10].

A first order reflectron lens consists of a series of annular fringing field electrodes and a reflector plate with increasing potentials, as shown in Fig. 4.10. As the ion enters the increasing field, it is retarded, reflected and then accelerated back out of the lens. The energy of the ion when it leaves the specimen is

$$\tfrac{1}{2}mv^2 = neV(1+\delta_e), \qquad 4.1$$

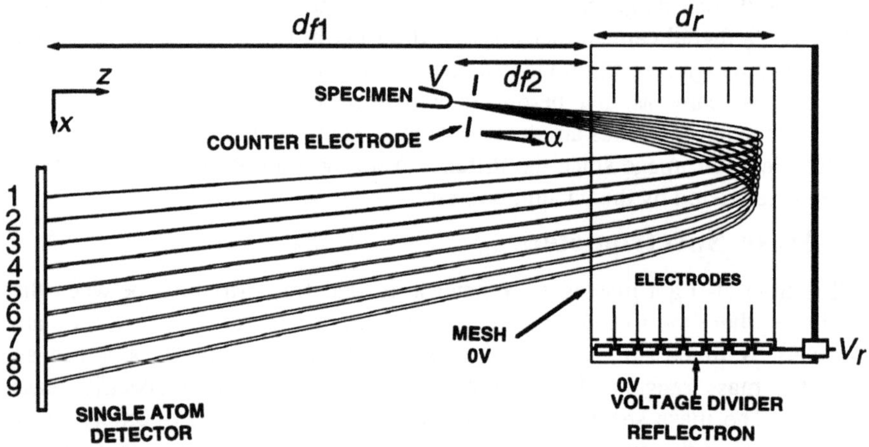

Fig. 4.10. An energy-compensating reflectron lens. A high tranparency mesh is used on the entrance and exit electrode to define the field.

where m is the mass of the ion, ne is the electronic charge, V is the effective total potential on the specimen, and δ_e is the energy variation. The axial velocity component, v_z, for an ion emitted at an angle α to the reflectron axis is given by [10]

$$v_z = v \cos\alpha = \sqrt{\frac{2neV}{m}}\sqrt{1+\delta_e}\cos\alpha. \qquad 4.2$$

When the ion enters the reflectron lens, it experiences a deceleration a given by

$$a = \frac{neV_r}{md_r} \qquad 4.3$$

where V_r is the voltage applied to the reflector plate and d_r is the distance traveled in the reflectron. Therefore the transit time in the reflectron lens is given by

$$t_r = 2\frac{v_z}{a} = 2\frac{md_r}{neV_r}\sqrt{\frac{2neV}{m}}\sqrt{1+\delta_e}\cos\alpha = 4\,d_r\frac{V_t}{V_r}\sqrt{\frac{m}{2neV_t}}\sqrt{1+\delta_e}\cos\alpha. \qquad 4.4$$

The transit time in the field free region outside the reflectron lens, t_f, is given by

$$t_f = \frac{d_f}{v_x} = d_f \sqrt{\frac{m}{2neV}}\, \frac{1}{\sqrt{1+\delta_e}\, \cos\alpha} \qquad 4.5$$

where d_f is the distance traveled in the field free region. This field free region is generally divided into two sections (d_{f2} and d_{f2}); the distance from the specimen to the reflectron lens and the distance from the reflectron lens to the single atom detector. The total flight time from the specimen to the single atom detector, t is the summation of eqns. 4.4 and 4.5, which after expansion of the $\sqrt{1+\delta_e}$ terms is given to a first approximation by

$$t = t_f + t_r \approx \frac{d_f}{\cos\alpha}\sqrt{\frac{m}{2eV}} \left(\begin{array}{l} 1+4\frac{d_r}{d_f}\frac{V}{V_r}\cos^2\alpha \\ -\frac{\delta_e}{2}\left[1-4\frac{d_r}{d_f}\frac{V}{V_r}\cos^2\alpha\right]+\frac{\delta_e^{2}}{8}\left[3+4\frac{d_r}{d_f}\frac{V}{V_r}\cos^2\alpha\right]+\cdots \end{array} \right) \qquad 4.6$$

Because the energy variation is small, typically 1%, the δ_e^2 and higher order terms can be disregarded. Therefore, the flight time of the ion will be independent of small variations in energy if the $\delta_e/2$ term of eqn. 4.6. is eliminated, i.e., if

$$4\frac{d_r}{d_f}\frac{V}{V_r}\cos^2\alpha = 1 \qquad 4.7$$

In practice, the distances in the field free region and the reflectron are set at a fixed ratio given by $d_f = 4d_r$. In the three-dimensional atom probe, the reflectron is usually positioned relatively close to the specimen to minimize the diameters of the apertures of the lens. The voltage applied to the final reflector plate electrode is maintained at a fixed ratio of the total applied voltage to the specimen. This ratio may be determined by accumulating mass spectra from a series of experiments with different V_r:V ratios. The mass resolving power is determined for each ratio from the full width at half maximum (FWHM) of the peaks in each mass spectrum. The ratio (typically ~1.05) that produces the maximum mass resolving power is then selected [11].

The voltage gradient on the individual anodes is defined by a matched series of UHV-compatible resistors rated for operation at high voltage. This voltage divider is generally incorporated into the vacuum system to reduce the number of high voltage feedthroughs. The typical value on these resistors (50-100 MΩ with ≤1% tolerance) is chosen to keep the current and consequently the heat generated in the resistors to a minimum so that the performance of the lens does not change with time. The voltage is generally supplied by a separate high voltage power unit although the same supply could be used for the standing voltage on the specimen and the reflectron lens with an appropriate variable voltage dividing resistor network. The entrance and exit electrode has a high transparency (~95%) mesh with a fine line spacing (~4 lines per mm) to ensure that the field is correctly terminated in a flat ground plane. This mesh reduces the detection efficiency of the collected ions by the square of the transparency since the ions pass through the mesh twice. Small aberrations due to the lensing effect of the mesh can be observed as a fine square pattern of lines on the field ion and single atom images. The effect of the reflectron lens on the spatial resolution is discussed in (§5.5.4).

In principle, significantly higher mass resolving power could be achieved with the use of a higher order reflectron lens. Higher mass resolving power ($m/\Delta m > 10{,}000$) is desired since it would enable the overlap between elements with common isobars and other overlapping species to be resolved (§5.4). However, these designs invariably include sections with different voltage gradients and therefore require additional meshes to terminate the sections that would further reduce the detection efficiency.

4.3.2 Primary detector

The single atom sensitive detector is located at the end of the mass spectrometer in the vacuum system. The active area of the detector and the flight distance determine the acceptance angle of the analysis area. The primary detector consists of two, or sometimes three, microchannel plates and a phosphor screen or other type of anode. The signal to stop the timing system is either taken from the rear or exit surface of the last microchannel plate or from the phosphor screen.

The initial designs featured circular microchannel plates (60-75 mm in diameter) whereas, the more recent designs have square (typically 10 by 10 cm) microchannel plates. The channels in the microchannel plate are cut at a small bias angle (typically ~8°) to the surface normal and are arranged in a chevron (+8°:-8°) or z-stack configuration. The channels typically have a length to diameter ratio of 60:1 or 80:1 (double thickness) so that saturation of the electron multiplication process is achieved. The microchannel plates that are normally used in single atom detectors are matched sets with similar resistances across the thickness of the plate so that appropriate voltages can be applied across each plate with a simple resistor chain type of voltage divider. Most designs feature a single power supply and a voltage dividing resistor network to supply the voltages for the microchannel plates and the phosphor screen or anode. The power supply to these detectors often features circuitry to slowly ramp the voltages to the operating levels over several seconds to minimize the possibility of discharges. Two configurations are in common use depending on whether the entrance surface to the microchannel plate or the anodes are at ground potential, as shown in Fig. 4.11. The electrons output from the microchannel plates are collected on either a phosphor screen or a multianode array. The phosphor screen is positioned as close as possible from the rear of the microchannel plate to minimize the spread of the electrons. In contrast, the multianode array in the tomographic atom probe is positioned a small distance (20 mm) from the microchannel plate so that the electron beam spreads out to cover several anodes. The type of phosphor is chosen for its response time and optical characteristics. Typical phosphors are P47 and P20.

The typical gain of a chevron microchannel plate configuration is 10^6-10^7 electrons for each incident ion and can be adjusted by altering the voltage applied across the microchannel plates. Saturation of a pair of microchannel plates occurs at a gain of approximately 10^8. At high gains, the number of electrons is sufficient to ionize gas atoms within the channel. These ions drift back down the channel where they may produce a second cascade of electrons. Since this cascade arrives at the anode at a later time than the original cascade, it may produce a spurious afterpulse in the timing electronics. This process is known as ion feedback and may be detected as a smaller amplitude peak on the high mass side of the peaks in the mass spectrum.

The typical detection efficiency of a chevron configuration is ~55-65%. The main reason for this low efficiency is that ions that strike the web region in between the channels are lost. To overcome this limitation, a negatively biased high transparency (98%) mesh can be incorporated in front of the microchannel plate to return the electrons produced by the incident ions into a channel [12,13], as shown in Fig. 4.12. With this arrangement, detection efficiencies of up to 95% can be achieved. Further improvement may be achieved by coating the front or entrance surface of the microchannel plate with a material with good secondary electron emission characteristics.

The electron multiplication process is not mass specific for ions with sufficient energy. However, low voltage specimens (i.e., less than ~2 kV) may suffer some loss in detection efficiency and therefore low operating voltages should be avoided.

The single atom detector should always be deactivated during the admission of image gas to the system to prevent damage to the expensive microchannel plates or phosphor screen if too much image gas is admitted.

4.3.3 Secondary detectors

Secondary detectors that are external to the vacuum system are used in the optical atom probe and its derivatives. These secondary detectors are optically focused onto the primary detector by either large aperture (f1-1.2) optical lenses or by fiber optic tapers. The primary reasons for the secondary detectors are to improve the detection efficiency when multiple ions strike the single

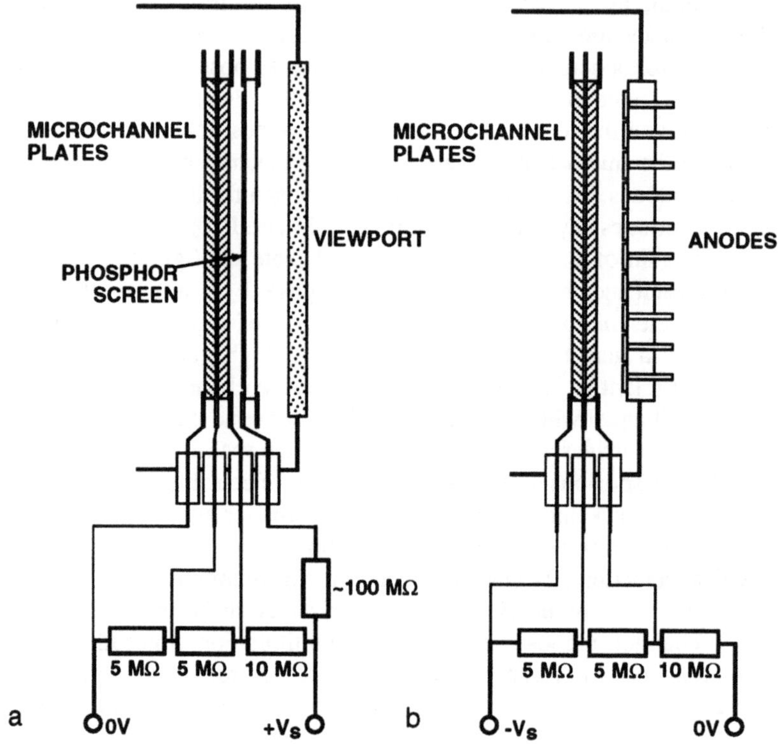

Fig. 4.11. Single atom detectors: a) phosphor screen and b) multianode array configuration.

atom detector on the same field evaporation pulse and to reduce the background noise level.

Two types of secondary detectors are in common use. The first is an image intensifier that was first used in these types of applications by Kellogg [14]. This image intensifier contains a photocathode to convert the optical signal (photons) into electrons, a microchannel plate to increase the number of electrons, and a phosphor screen to convert the electrons back into an optical signal. The spacings between these three plates are small so the position information is not significantly degraded by the image intensifier. A CCD video camera and a second lens or fiber optic taper are used to record the resulting image on the phosphor screen. In order to attain sufficient spatial resolution, the size of the pixel array in the CCD camera is at least ~512 by ~512. The CCD camera is connected to a frame grabber that is usually

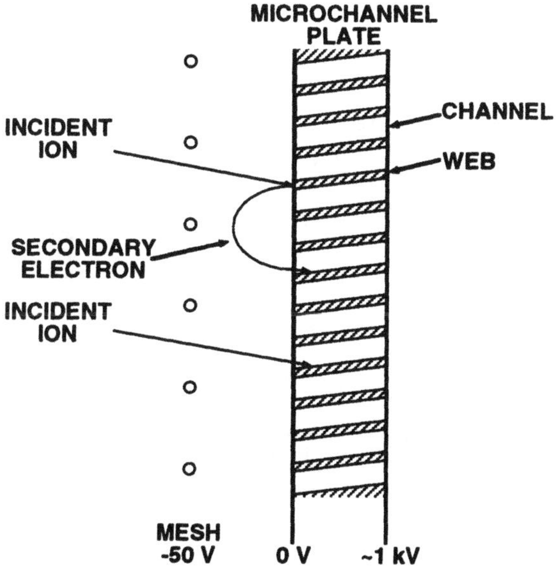

Fig. 4.12. Single atom detector incorporating a mesh to improve the detection efficiency by returning the secondary electrons emitted from the interchannel regions.

incorporated directly in the control computer. This combination provides a digital image of the position and intensity of the spots on the detector. The frame rate of the camera is another of the primary factors that controls the data accumulation rate of the instrument.

The second type of secondary detector is similar except that the phosphor screen is replaced by a multianode array. The number of anodes in the multianode array is ~8 by ~10 and is a compromise between the number of timing channels required, the effective size of each anode and the dead space between the anodes. In principle, a larger number of anodes should provide a better correlation between the spots on the camera and the hits on the primary detector since there is a lower probability that two ions will strike one anode. However, this increases the dead space between the anodes and the number of timing channels required.

If two secondary detectors are used, a beam splitter such as a front silvered semi-transparent mirror set at 45° to the main optical path is required. A schematic diagram of a dual configuration is shown in Fig. 4.13. The ratio of reflected and transmitted signals can be adjusted to suit the sensitivity of the secondary detectors. The CCD camera detector is normally positioned to

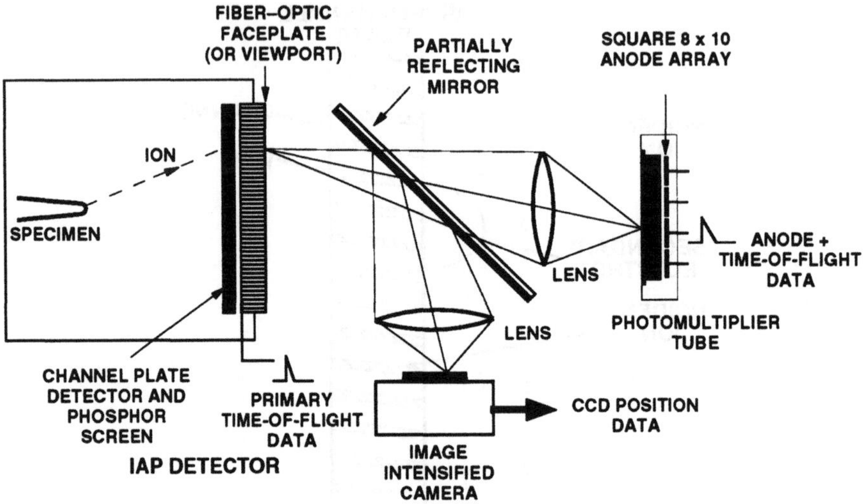

Fig. 4.13. Schematic diagram of an optical position-sensitive atom probe with a dual secondary detection system featuring a CCD camera and a multianode array.

receive the higher quality image from the reflected surface of the mirror, because it is used to estimate the spatial position.

In both variants, the photocathode can be activated in a similar manner to a high speed shutter by applying a gate pulse only during the time when genuine ions strike the primary detector. These time-gated image intensifiers significantly decrease the noise level. These secondary detectors must only be operated in a light tight environment.

4.3.4 Timing system

The flight time of the ion from the specimen to the primary detector is measured by a time-to-digital converter (TDC). The timing system is started by a signal from the high voltage pulse used to field evaporate the ions (§4.2.3) and stopped by an amplified signal from the primary detector (§4.3.3). The stop signals from the primary detector are normally amplified in a preamplifier positioned as close to the detector as possible and then shaped in a discriminator prior to the time-to-digital converter. A short delay (100-300 ns) in activating the time-to-digital converter is normally required to prevent false stop signals that are generated from the high voltage pulse. This delay is generated by either a length of coaxial cable or a delay gate generator module. In instruments with multiple anodes, an individual preamplifier and timing channel is used for each anode. The type of preamplifiers used on the primary and secondary detectors are generally not identical and therefore have different transmission times. The time-to-digital converters are usually capable of detecting multiple (up to 8 or 16) ions per field evaporation pulse. In some time-to-digital converters, such as the LeCroy™ 2277 or 3377 modules, the width of the stop pulse may be determined from the times of the positive and negative crossings of the threshold set on the internal discriminator. This additional information may be used to estimate the number of ions striking the detector at the same time as the pulse width increases with the number of ions. Since there are significant variations in amplitudes and widths of the stop pulses, this estimate is only approximate. The time-to-digital converter measures the time between the start and stop signals at the module rather than when the ion leaves the specimen and strikes the detector. Therefore, a correction is required to account for the transmission times in the preamplifiers, discriminators and the connecting cables. These values can be estimated with an oscilloscope or from calibration experiments (§5.4).

The typical flight times of ions as a function of applied voltage and mass-to-charge ratio are shown in Fig. 4.14 for a flight distance of 620 mm. In order to measure the flight time for all mass-to-charge ratios up to 500 u at all applied voltages, a timing period of ~30 μs is required for a mass spectrometer with a flight distance of 620 mm. However, it is extremely rare to observe ions with mass-to-charge ratios of over 209 u (Bi^+) in a voltage pulsed atom probe and the vast majority of ions have mass-to-charge ratios of less than 100 u. The bin size or width of a timing channel in the time-to-digital converters is 0.5 or 1 ns. The variation of the mass-to-charge ratio resolution of a 1 ns wide bin with applied voltage and mass-to-charge ratio is shown in Fig. 4.15. Due to the time to mass conversion, there are more time bins per mass-to-charge ratio bins at lower masses.

4.3.5 Computer system and interface

The timing system and the associated electronics are controlled by a microcomputer, as shown in Fig. 4.16. The computer is connected to the electronics with the use of a high speed interface. The data transfer rate over this interface is another of the primary factors that controls the data accumulation rate of the instrument.

This computer is used to read and set the various components described in the previous sections, to control the experiment, and to provide some on-line information to the operator during the experiment. On-line displays may be divided into two groups: diagnostic displays to ensure correct operation of the instrument and specimen-related displays to inform the operator as to the progress of the experiment. Some examples of diagnostic displays include a) a status display of the current voltages on the specimen and lens, the

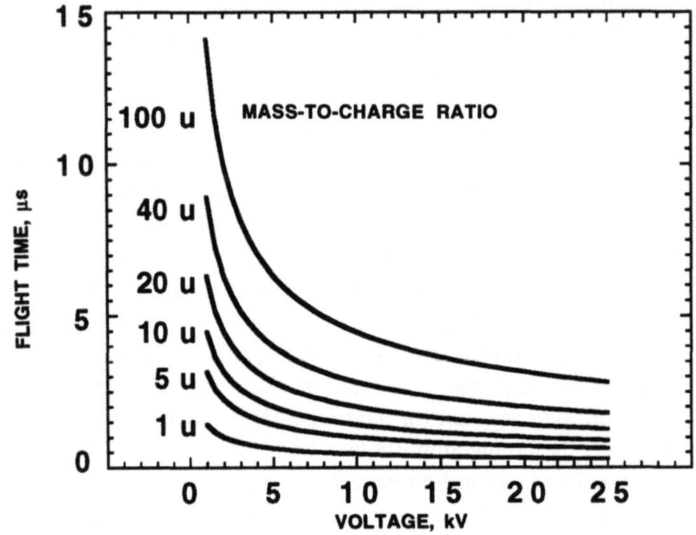

Fig. 4.14. Flight times of ions at different applied voltages for a flight distance of 620 mm.

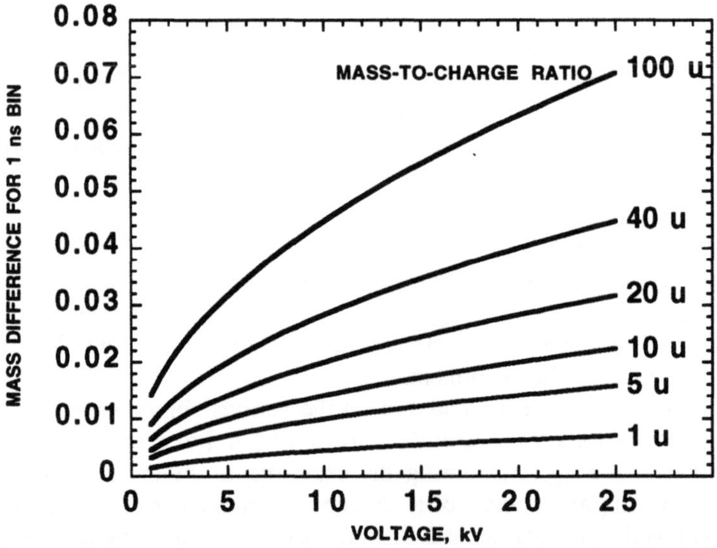

Fig. 4.15. The variation in the mass-to-charge ratio resolution for a 1 ns timing bin as a function of applied voltage.

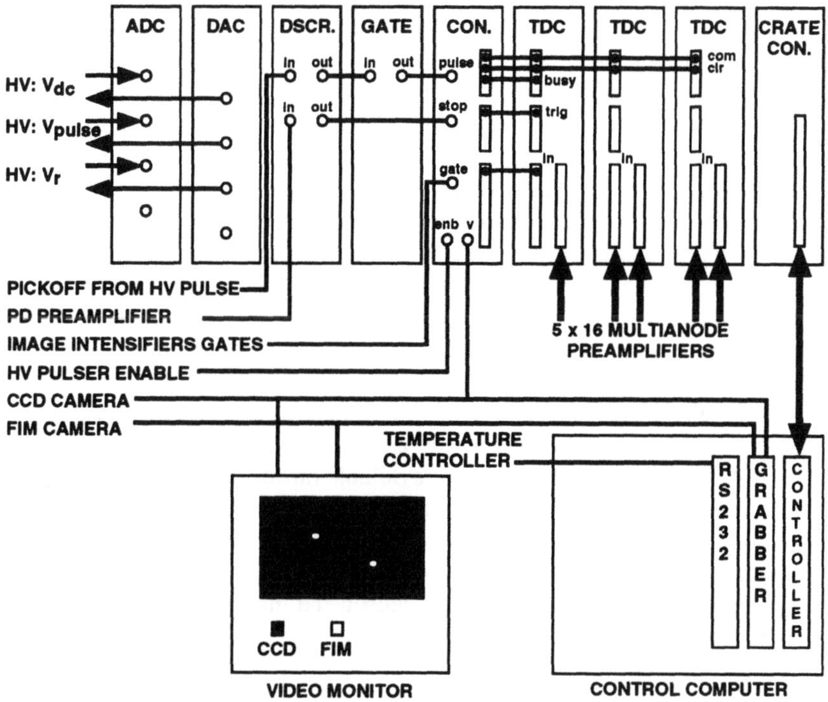

Fig. 4.16. The timing system used to determine the flight times of the ions and to control the voltages on the specimen and reflectron lens.

number of field evaporation pulses required to remove the last ion, the average evaporation rate, and the time, mass-to-charge ratio and identity of each ion, b) the distributions of the number of ions detected on each field evaporation pulse on the primary detector, c) the number of hits on each anode, and d) the intensity distributions of the signals recorded on the CCD camera. Some examples of specimen-related displays include the mass spectra, simple composition profiles, two dimensional dot maps of the impact positions on the detector, and the accumulated number of ions collected as a function of applied voltage on the specimen.

Most computer systems are designed to permit unattended operation of the instrument for extended periods. The computer automatically increases and decreases the voltages on the specimen to maintain a constant average evaporation rate (or more accurately an average ion collection rate) set by the operator. Typically, the voltages are increased after a set number of pulses during which no ions are collected and when the average number of ions collected per pulse drops below a set level. In addition, if the average number of ions per pulse exceeds another set level, the voltages are decreased. These average ion collection rates are typically calculated over the preceeding (~20) ions. Although fixed parameters are used in some programs, it is more desirable to maintain a constant field on the specimen and therefore, these parameters are adjusted to take into account the change in radius and hence the field on the specimen during the experiment. The computer system is also capable of terminating the experiment in the case of a specimen failure or when the specimen voltage limit is reached. In some instruments, systems are incorporated to detect sudden increases in pressure in the vacuum system so that the delicate components such as the detector used to record the field ion image, the single atom detectors and other electronics can be deactivated before damage occurs. Systems that enable the atom probe to be remotely monitored and controlled over the Internet are also available on some instruments.

Since analysis of the three-dimensional data is computationally intense, another higher power computer is generally used to reconstruct and visualize the three-dimensional data after the experiment has been completed. Details of the types of analyses that can be performed on the three-dimensional data are given in Chapter 6.

4.4 Instruments

Several different types of three-dimensional atom probes have been developed. Although there are differences in the overall design of these different instruments, the primary distinguishing feature between these instruments is the type of single atom detector and the method used to determine the coordinates of the impact of the ions. All these single atom detectors use different approaches to determine the positions of the electron clouds that are generated in the microchannel plates. The unique features of these single atom detectors and their advantages and limitations are discussed in this section.

4.4.1 Position-sensitive atom probe

The position-sensitive atom probe uses a wedge-and-strip anode to encode the position of the ion's impact on the detector [15-17]. The wedge-and-strip anode that is used in this instrument has three electrodes, as shown in Fig. 4.17. The X electrode is an interconnected equally-spaced series of identical triangular-shaped wedges. The Y electrode is an interconnected series of rectangular strips which progressively increase in thickness from one side of the detector to the other. The Z electrode is the region in between these other two electrodes. The X_a and Y_a positions of each ion striking the detector are

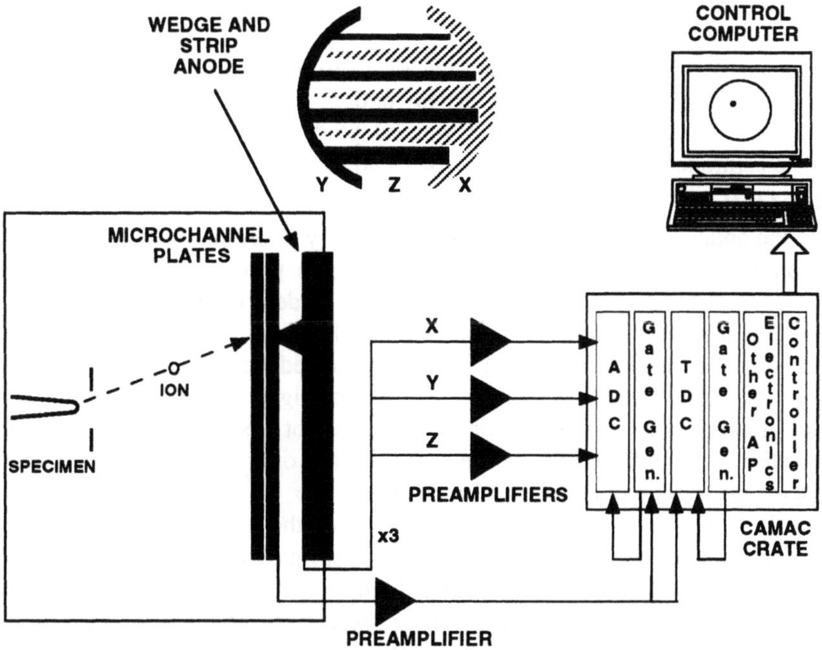

Fig. 4.17. Schematic diagram of the position-sensitive atom probe. This instrument features a wedge-and-strip detector to determine the impact position of ions on the detector.

determined from the relative charges, Q_X, Q_Y, Q_Z, measured on the X, Y and Z electrodes, respectively

$$X_a = 2Q_X/(Q_X+Q_Y+Q_Z) \text{ and } Y_a = 2Q_Y/(Q_X+Q_Y+Q_Z). \qquad 4.8$$

The typical accelerating voltage applied between the output face of the microchannel plate and the anode (200-500 V) is much lower than that normally used (>3 kV) in a typical single atom detector. This lower voltage ensures that the electron cloud produced by the microchannel plate diverges sufficiently to cover a few adjacent wedges and strips.

This type of detector provides accurate data only when a single ion strikes the detector on any field evaporation pulse. If multiple ions with similar flight times are field evaporated on the same pulse, an average position of the multiple ion impacts is obtained. To minimize this problem, the experimental conditions are adjusted so that the probability of multiple field evaporation events is low.

An alternative variant of this type of instrument with a more elaborate wedge-on-wedge (WoW) detector is under development by Kelly and coworkers [18-20]. This approach was adopted to allow multiple ions to be detected simultaneously through the use of additional electrodes.

4.4.2 Optical atom probe

In the optical atom probe [20,21], the x and y coordinates of the impact of the ion are determined from the image on the phosphor screen of the primary detector with a sensitive high speed CCD video camera, as shown in Fig. 4.18. The time-gated image intensifier tube (§4.3.3) between the primary detector and the camera acts as a fast shutter and reduces the background noise level. The CCD camera is interfaced to an intelligent CAMAC module or a frame grabber in the control computer. It is not necessary to transfer the intensities of the entire image to the computer but only a few valid pixels thus reducing the time required to acquire the data.

In the optical atom probe, the flight time of the ion is measured from a signal taken from the phosphor screen on the primary detector. When a valid flight time is detected, the intensities of the pixels that are above a selected threshold are encoded and transferred to the control computer. The x and y position of the pixel with the highest intensity is then determined. A typical map of the intensity distribution from a single ion impact on the detector is shown in Fig. 4.19. The lateral spatial resolution of the detector is defined by the effective size of a pixel in the CCD array. The impact position is given by

$$X_a = \frac{x}{x_{\max}} \quad \text{and} \quad Y_a = \frac{y}{y_{\max}} \tag{4.9}$$

where x and y are the coordinates of the pixel with the highest intensity and $x_{\max}$ and $y_{\max}$ are the maximum extents of the CCD array. Typical pixel resolutions are 1 part in 256 to 512. In principle, an interpolation scheme could be adopted to obtain sub-pixel resolution. For example, the center of intensity $(\bar{x}, \bar{y})$ of a spot may be given for isolated spots by

$$\bar{x} = \frac{\sum_{i=1}^{n} x_i I_i}{\sum_{i=1}^{n} I_i} \quad \text{and} \quad \bar{y} = \frac{\sum_{i=1}^{n} y_i I_i}{\sum_{i=1}^{n} I_i} \tag{4.10}$$

where I_i are the pixel intensities, and n is the number of pixels with intensities above a specified threshold that make up the spot.

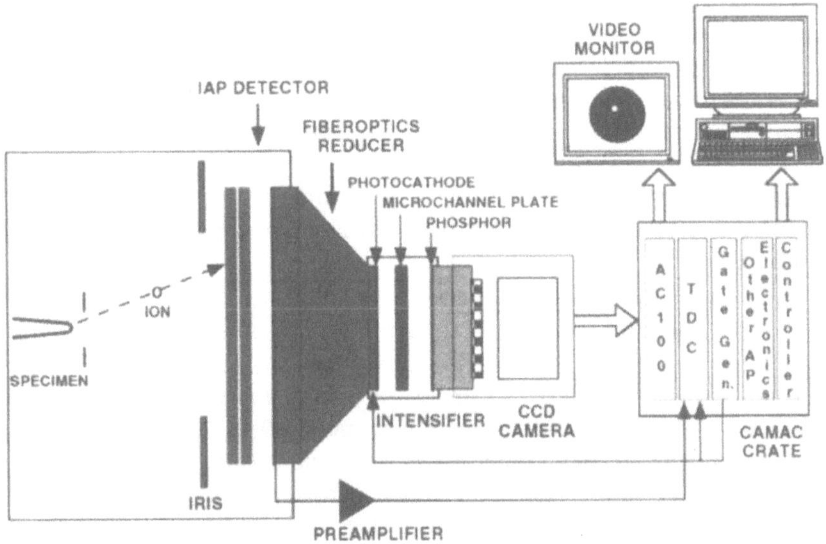

Fig. 4.18. The optical atom probe features an external time-gated image intensifier and CCD camera to encode the positions of the ions striking the primary detector.

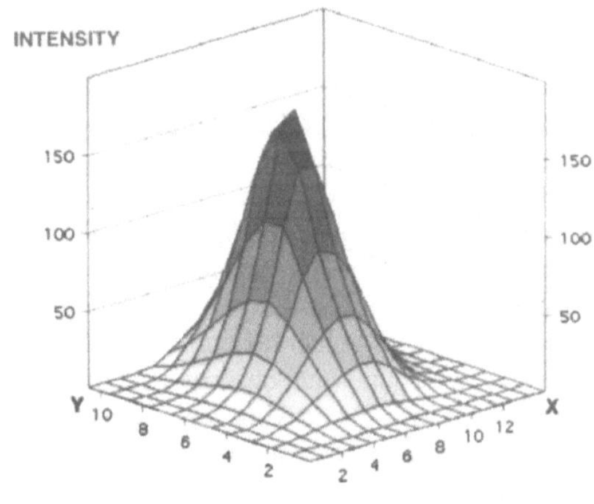

Fig. 4.19. The intensity distribution measured from a CCD camera of a single ion striking an optical atom probe detector.

The original optical atom probe suffers from the same limitation as the position-sensitive atom probe in that only one ion should be permitted to strike the primary detector for each field evaporation pulse. When more than one ion strikes the primary detector simultaneously, it is not possible to identify which flight time is associated with which impact position, although both positions and times are measured accurately. However, if both ions can be assigned to the same element (i.e., the same mass or different isotopes or charge states of the same element) then it is reasonable to assign that element to both measured positions.

4.4.3 Tomographic atom probe

The tomographic atom probe (TAP) was the first instrument to use a multianode array and a parallel multichannel timing system [23-25]. The detector consists of a z-stack of three 105 mm diameter circular microchannel plates, an 80 by 80 mm entrance window and a 10 x 10 multianode array, as shown in Fig. 4.20. The distance (20 mm) and relative voltage between the microchannel plates and anodes are critical parameters in this design. Since the anodes are operated at ground, the front face of the first microchannel plate is held at a negative voltage. Therefore, to preserve a field-free drift zone in the mass spectrometer, a grounded 95% transparent grid is placed 8 mm in front of the first microchannel plate. A second grid positioned between the grounded grid and the front microchannel plate and held at a slightly more negative voltage than the microchannel plate improves the detection efficiency of the detector (§4.3.2). A low noise preamplifier is connected to each of the 96 anodes (the 4 corner anodes of the 10 x 10 array are not used). Each preamplifier is connected to a parallel stack of up to eight analogue-to-digital (ADC) FASTBUS modules each of which has 96 input channels (i.e., one channel for each anode). The X and Y spatial coordinates of each impact on the anode array are determined by interpolating the relative charge measured on the adjacent anodes. The time-of-flight signals are taken from the rear of the third microchannel plate and are used to trigger the successive ADC modules for multiple ions, as shown in Fig. 4.20. This design permits up to eight time windows to be observed for each field evaporation pulse.

The lateral spatial resolution of this multianode detector is a function of the accuracy of the interpolation routines used to calculate the impact position, the number of contiguous anodes covered by the charge cloud, and the number and proximity of ions striking the detector in the same window. In addition, some assumptions are required for the shape and spatial distribution of the charge within the charge cloud and their variations with the total charge produced by each ion. If charge is detected on one anode, the precise impact position is somewhere near the center of the anode. Similarly, if charge is detected on two adjacent anodes, the impact position is known accurately in

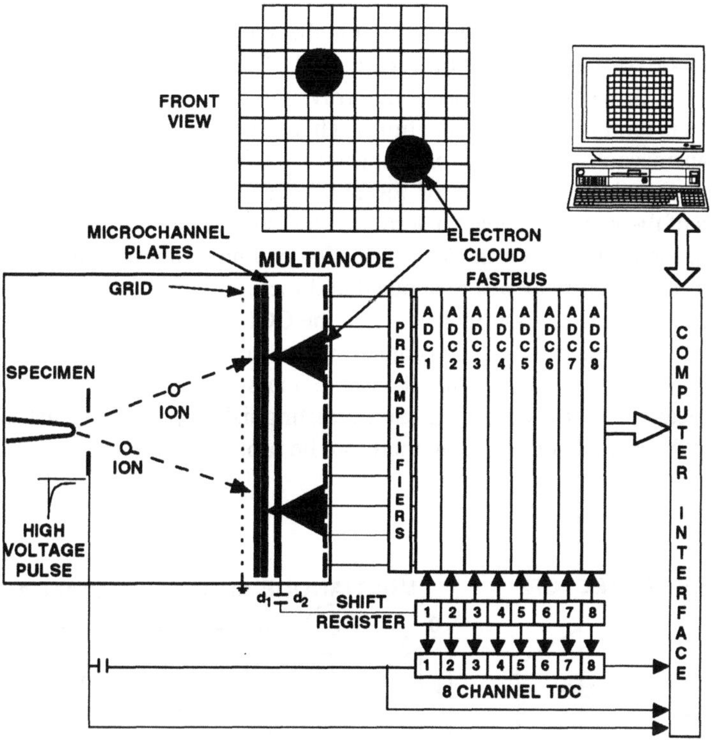

Fig. 4.20. The tomographic atom probe features a 10 by 10 array of anodes and a parallel timing system to determine the positions of the ions striking the detector.

only one of two directions. However, these two special cases only account for a small fraction of the ions collected. Under the optimum operating conditions when the charge cloud covers three or more anodes and there are no multiple interpenetrating impacts in the same window, the position of the impact can be determined by a method based on charge centroiding [26-29]. This method has a spatial resolution that is significantly better than the size of the anodes.

4.4.4 Optical position-sensitive atom probe

The optical position-sensitive atom probe is a derivative of the optical atom probe [30] and the detector bears no relationship to the wedge-and-strip detector used in the original position-sensitive atom probe. This variant may be described as an optical atom probe with an additional multianode secondary detector, as shown in Fig. 4.13. This dual configuration enables the time-of-flight information from the primary detector (t_1 and t_2) to be linked to the time-of-flight information from the anodes on the secondary detector (t_1':anode 0 and t_2':anode 3), as shown in Fig. 4.21 [31]. The flight times on the primary detector and those on the multianode array will not be identical and a small correction is required to match the times. This discrepancy is due to differences in the transmission times in the preamplifiers, cables and the multianode array. Then that area imaged by one of the anodes can be assigned to one of the positions of the primary impact determined by the CCD camera with the use of a look up table, as shown in Fig. 4.21. For example, the coordinate x_a,y_a on the CCD camera corresponds to anode 0 and the coordinate x_b, y_b corresponds to anode 3. Therefore, coordinate x_a,y_a on the CCD camera corresponds to time t_1 on the primary detector and coordinate x_b, y_b corresponds to time t_2. In this method, two or more ions with the same mass-to-charge ratio striking the primary detector at the same time can be distinguished providing that they did not strike the same anode and can therefore be properly assigned.

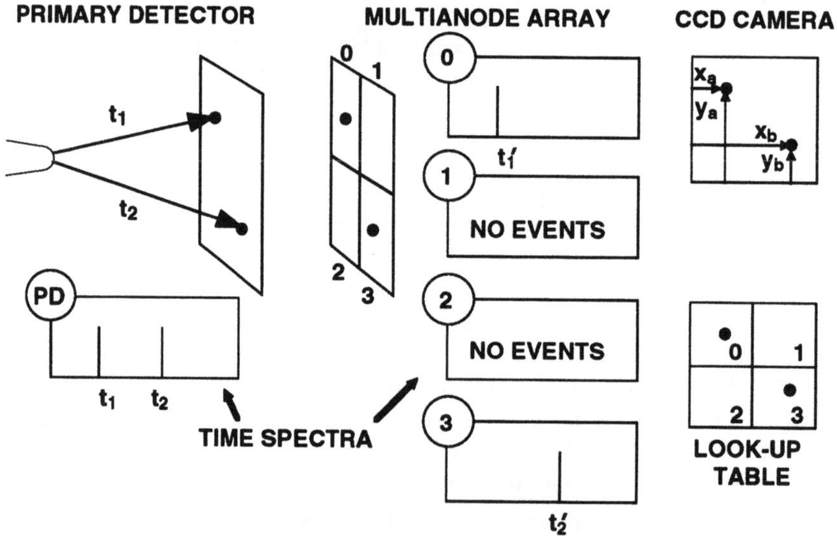

Fig. 4.21. Process to assign the flight times from the primary detector to the positions recorded on the CCD camera with the use of the flight times and anode positions from the multianode array.

4.4.5 Optical tomographic atom probe

Another enhanced variant of the optical atom probe that uses a CCD camera for positioning the ions is the optical tomographic atom probe [32]. A schematic diagram of this instrument is shown in Fig. 4.22. In this instrument, the phosphor screen on the primary detector is divided into 16 transparent strip anodes each connected to a preamplifier and a time-to-digital converter for the time-of flight information. This arrangement essentially combines a dual detector configuration into a single detector and eliminates the need for a beam splitter. It also has the advantage that the image area and the anode array are physically identical. Therefore, there are fewer registration problems between the primary and secondary detectors.

To achieve a similar functionality, an external image intensifier that incorporates a 16 x 2 array of transparent anodes in place of the normal phosphor screen has also been proposed [33]. This configuration has the advantage that it can be incorporated into existing imaging atom probes and optical atom probes without modification to the vacuum system.

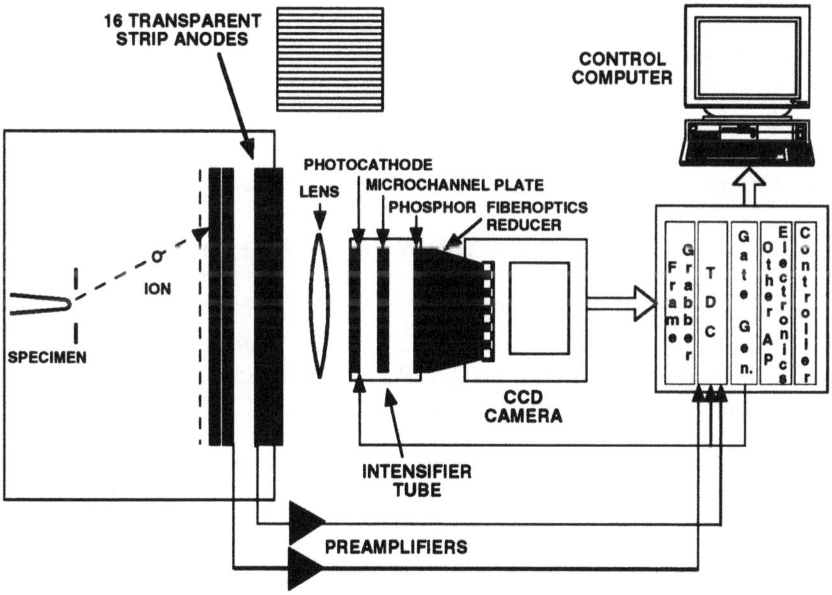

Fig. 4.22. The optical tomographic atom probe features a transparent multianode to detemine the positions of the ions striking the detector.

References

1. M. K. Miller and G. D. W. Smith, Atom Probe Microanalysis: Principles and Applications to Materials Problems, Material Research Society, Pittsburgh PA, 1989.
2. T. J. Godfrey, R. M. Cripps and G. D. W. Smith, *J. Phys. E,* **10** (1976) 329.
3. A. R. Waugh, *J. Phys. E.*, **14** (1981) 615.
4. G. L. Kellogg and T. T. Tsong, *J. Appl. Phys.*, **51** (1980) 1184.
5. E. W. Müller and S. V. Krishnaswamy, *Rev. Sci. Instrum.*, **45** (1974) 1053.
6. W. P. Poschenreider, *Int. J. Mass Spectrom. Ion Phys.*, **9** (1972) 83.
7. A. N. Kudryavtecm, N. V. Nikonenko, B. M. Dubenskii and D. V Shmikk, *Sov. Tech. Phys. Lett.*, **15** (1989) 261.
8. W. Drachsel, L. v. Alvensleben and A. J. Melmed, *J. de Phys.*, **50-C8** (1989) 541.
9. P. P. Camus and A. J. Melmed, *Surf. Sci.*, **246** (1991) 450.
10. A. Cerezo, T. J. Godfrey, S. J. Sijbrandij, G. D. W. Smith and P. J. Warren, *Rev. Sci. Instrum.*, **69** (1988) 1.
11. S. J. Sijbrandij, A. Cerezo, T. J. Godfrey and G.D.W. Smith, *Appl. Surf. Sci.*, **94/95** (1996) 428.
12. B. Deconihout, P. Gerard , M. Bouet and A. Bostel, *Appl. Surf. Sci.*, **94/95** (1996) 422.
13. S. J. Sijbrandij, A. Cerezo, B. Deconihout, T. J. Godfrey and G.D.W. Smith, *J. de Phys. IV*, **6-C5** (1996) 297.
14. G. L. Kellogg, *Rev. Sci. Instrum.*, **58** (1987) 38.
15. A. Cerezo, T. J. Godfrey and G. D. W. Smith, *Rev. Sci. Instrum.*, **59** (1988) 862.
16. A.Cerezo, T. J. Godfrey, and G.D.W. Smith, *J. de Phys.*, **49-C6** (1988) 25.
17. A. Cerezo, *Vacuum*, **42** (1991) 605.
18. T. F. Kelly, J. J. McCarthy, and D. C. Mancini, High-repetition rate position sensitive atom probe, U.S. Pat. No. 5061850, (1991).
19. T. F. Kelly, D. C. Mancini, J. J. McCarthy, and N. A. Zreiba, *Surf. Sci.*, **266** (1991) 396.
20. P. P. Camus, D. J. Larson, L. M. Holzman and T. F. Kelly, *J. de Phys. IV*, **6-C5** (1996) 291.
21. M. K. Miller, *Surf. Sci.*, **246** (1991) 428.
22. M. K. Miller, *Surf. Sci.*, **266** (1992) 494.

23. A. Bostel, D. Blavette, A. Menand, and J. M. Sarrau, *J. de Phys.*, **50-C8** (1989) 501.
24. D. Blavette, B. Deconihout, A. Bostel, J. M. Sarrau, M. Bouet and A. Menand, *Rev. Sci. Instrum.*, **64** (1993) 2911.
25. B. Deconihout, A. Bostel, A. Menand, J. M. Sarrau, M. Bouet, S. Chambreland and D. Blavette, *Appl. Surf. Sci.*, **67** (1993) 444.
26. B. Deconihout, S. Chambreland, and D. Blavette, *Adv. Mater.*, **6** (1994) 695.
27. B. Deconihout, A. Bostel, M. Bouet, J. M. Sarrau, P. Bas and D. Blavette, *Appl. Surf. Sci.*, **87/88** (1995) 428.
28. P. Bas, A. Bostel, B. Deconihout, and D. Blavette, *Appl. Surf. Sci.*, **87/88** (1995) 298.
29. P. Bas, A. Bostel, G. Grancher, B. Deconihout and D. Blavette, *Appl. Surf. Sci.*, **94/95** (1996) 442.
30. A. Cerezo, T. J. Godfrey, J. M. Hyde, S. J. Sijbrandij, and G. D. W. Smith, *Appl. Surf. Sci.*, **76/77** (1994) 374.
31. A. Cerezo, J. M. Hyde, S. J. Sijbrandij and G. D. W. Smith, *Appl. Surf. Sci.*, **94/95** (1996) 457.
32. B. Deconihout, L. Renaud, G. Da Costa, M. Bouet, A. Bostel and D. Blavette, *Ultramicroscopy*, **73** (1998) 253.
33. M. K. Miller, *Microsc. Microanal.*, **4** suppl 2 (1998) 80.

Chapter 5

Experimental Factors

The steps that are required to perform an atom probe analysis and the experimental factors that affect the performance of the instrument are described in this chapter. Since atom probe tomography is a destructive technique, it is essential that these factors be taken into consideration before starting an atom probe experiment in order to obtain quantitative data.

5.1 Atom Probe Analysis Procedure

The normal sequence of steps to perform an atom probe experiment is as follows:

a) Perform the steps described in §3.1.1 to form a field ion image. It is generally necessary to turn off the ion gauge and inhibit the titanium sublimation pump on the analysis chamber before proceeding.
b) Field evaporate the specimen until a full, uniform and clean surface is obtained. If a specific feature of interest is to be analyzed, continue to field evaporate the specimen until that feature is visible in the field ion image. Set the voltage on the specimen to the best image voltage and record the field ion image. Voltages between ~5 and ~12 kV are generally suitable to begin an analysis. The lowest starting voltage (~2 kV) is dictated by the non-uniformity of the detection efficiency of the single atom detector for different types of ions. The upper limit for the voltage range (~15 to ~20 kV) is usually defined by the maximum voltage that can be generated by the high voltage pulser to maintain a constant pulse fraction (§4.2.3).
c) Start the program to control the instrument. It is normal to enter parameters such as the data file name, the specimen identification

information, an existing range file name or create new range windows (§5.4.6), and the experimental parameters (pulse fraction, minimum and maximum permissible evaporation rate, etc.).

Some of the operations described in the following sections may be performed automatically by the control program and some may not be applicable to all types of three-dimensional atom probes.

d) Deactivate the field ion detector and activate the single atom detector in low gain or field ion mode.
e) Activate the reflectron lens and increase the applied voltage on the lens until it slightly exceeds the standing voltage on the specimen.
f) Select a position on the specimen surface for analysis (as described in §5.2) and move the specimen until the field ion image of that region is observed on the single atom detector. Record both angles of rotation from the specimen axis. It is advisable to also record the field ion image of the starting position on the single atom detector. The position of the specimen should not be adjusted during the collection of data.
g) Remove the image gas from the system. (Unlike the classical atom probe, all analyses in three-dimensional atom probes are performed at the base pressure of the instrument).
h) Ensure all sources of light and electron emission in the analysis chamber that would effect the single atom detector are removed or deactivated.
i) When the vacuum in the analysis chamber has reached a safe pressure (typically $<5 \times 10^{-10}$ mbar), set the single atom detector to the high gain or single atom sensitive condition. Activate secondary detectors, preamplifiers and the timing system.
j) Decrease the standing voltage on the specimen by at least the voltage that will be applied to specimen by the high voltage pulse to prevent premature failure on the application of the high voltage pulse.
k) Activate the high voltage pulser and select the appropriate pulse fraction (§5.3) and pulse repetition rate.
l) Select display options on the control computer (§4.3.5).
m) Start applying high voltage pulses to the specimen to begin the analysis. Either manually or automatically increase the voltages on the specimen and lens as required to maintain the selected average evaporation rate. Some operators prefer to manually control the voltage increase during the initial stages of the experiment and then switch to automatic operation when a steady ion collection rate is achieved.
n) Continue to analyze the specimen until a desired number of atoms are collected, the selected feature has been analyzed, the standing or pulse

voltage limits of the instrument are reached, or the specimen fails. This step may take several hours. Record the final standing and pulse voltages on the specimen for use with the calibration of the radius of the specimen (step r and §5.5). Stop the pulses from the high voltage pulser and deactivate the single atom detector, secondary detectors, preamplifiers, timing system, and high voltage pulser.

o) If the specimen has not failed, decrease the voltage on the specimen slightly to prevent field evaporation in the presence of the image gas and admit image gas to a pressure of ~1 to 5 x 10^{-5} mbar. Either activate the single atom detector in the low gain condition or move the specimen to the field ion detector position and activate the field ion detector. Adjust the standing voltage on the specimen to the best image voltage and record the voltage and a field ion image. Increase the voltage on the specimen until field evaporation just initiates and record the voltage. If the specimen failed, omit this step and step r.

p) Deactivate the standing voltage power supply, reflectron lens, single atom detector and field ion detector. Remove the image gas. Return the specimen to the transfer position. If necessary activate ion gauges and remove the inhibition on the titanium sublimation pump.

q) Terminate the control program. Make an archive copy of data file. Transfer a copy of data file to analysis computer.

r) Transfer the specimen to a transmission electron microscope in order to determine the end radius of the specimen. This parameter is required for the accurate reconstruction of the three-dimensional data (§5.5).

The analysis of surface layers requires some stages on this procedure to be omitted. Specifically, no field ion microscopy and no field evaporation of the specimen are performed prior to analysis and the standing and pulse voltages on the specimen are ramped up from lowest voltages required to operate the timing system.

5.2 Volume of Analysis and Geometrical Considerations

In order to perform a successful analysis in a three-dimensional atom probe, some awareness of the volume that may be analyzed and the number of atoms contained in a particular feature are required so that the optimum experimental approach is adopted.

The active area of the single atom detector defines the area of the specimen that may be analyzed. A comparison of the area of analysis to the field of view of the field ion image is shown in Fig. 5.1. As the experiment progresses, the specimen blunts due to the taper angle of the specimen and therefore, the magnification of the image decreases and the extent of the area

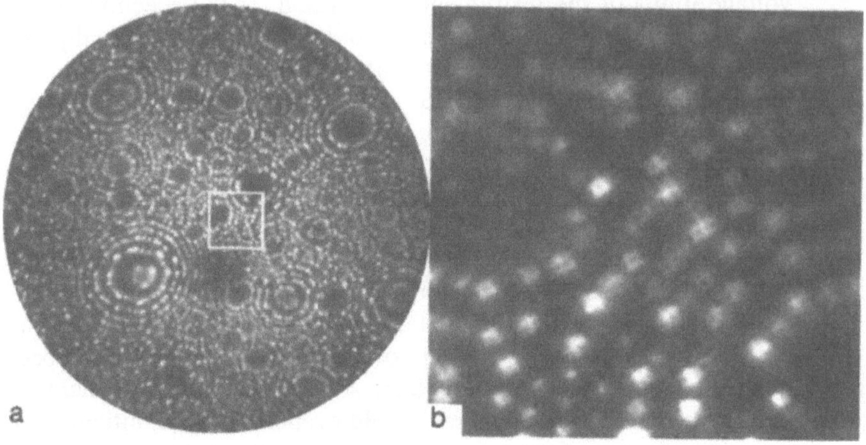

Fig. 5.1. Comparison of the field of views of the field ion image and the area of the specimen on the single atom detector.

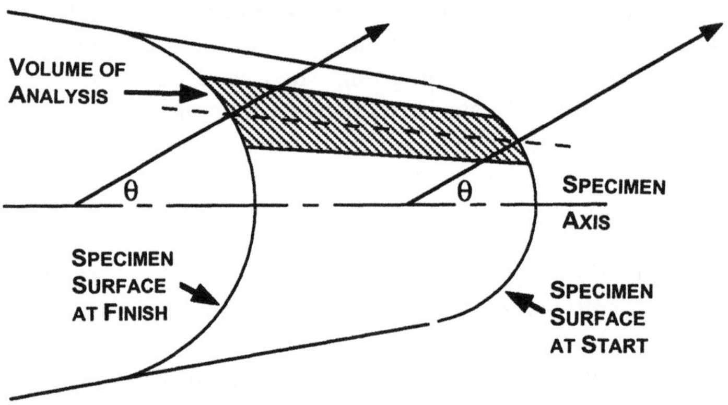

Fig. 5.2. Orientation of the volume of analysis with the axis of the specimen. The volume analyzed makes a small angle with the axis of the specimen.

analyzed increases. This process creates an analysis volume with the shape of a tapered cone with either a square or circular cross section depending on the geometry of the single atom detector. The ends of the analysis volume exhibit the curvature of a surface of a sphere and their precise shape depends on the angles of rotation from the axis of the specimen. The orientation of the main axis of the analyzed volume is shown schematically in Fig. 5.2. If the center of the analyzed volume is positioned at an angle θ to the specimen axis at the start of the experiment, the angle will not change during the analysis. Therefore, the axis of the analyzed volume is along the straight line connecting these points. It should be emphasized that the axis of the analyzed volume is inclined at a small angle to the axis of the specimen and not perpendicular to the surface [1,2].

If the number density of the features of interest such as a precipitate is sufficiently high, typically $>\sim 10^{23}$ m^{-3}, the analysis may be performed starting at any location on the specimen surface. In this type of random area analysis, it is customary to position the specimen such that the central axis of the specimen is aligned with the center of the single atom detector. It is advantageous to position the specimen close to but not on a major pole near the center of the specimen so that the interatomic spacing between the planes may be used to calibrate the scale in the z-direction. Single crystal specimens oriented with a major crystallographic direction along the axis of the specimen and materials with predominant wire textures are particularly useful for this approach. It is advisable not to select the center of planes or certain zone axes for analysis since the trajectory aberrations are most severe in these locations. The use of the central portion of the specimen also avoids any edge effects and minimizes the possibility of artifacts from the specimen preparation process.

If the number density of features is too low or the feature of interest is a boundary, interphase interface, dislocation or other planar feature, then the specimen is field evaporated until that feature of interest is visible in the field ion image. The specimen is then positioned so that the feature of interest is visible on the single atom detector. Some of the feature will have been field evaporated prior to the start of the analysis but such pre-analysis evaporation should be kept to a minimum. This is particularly important in the case of small precipitates as the number of atoms available for analysis is limited and therefore field evaporation of the precipitate should be minimized.

The selection of the initial analysis position must also take into account how the analyzed volume will intersect the feature as the analysis proceeds. Therefore, a precipitate that is smaller than the projected area of analysis should be positioned in the center of the single atom detector. In the case of a boundary, interface or larger precipitate, the selection of the starting position depends on the orientation of the boundary plane or interface with the axis of the specimen. The projected change in position of a boundary is discussed in §3.2.6. If the boundary plane is along the specimen axis, the boundary should be positioned in the center of the detector and the analyzed volume should contain the boundary along the direction of analysis. If the boundary plane is perpendicular to or inclined at an angle to the specimen axis, the image of the boundary should be positioned close to one edge of the single atom detector such that the center of the detector is on the inside of the circular image of the boundary plane. In this case, the progress of the experiment should be monitored in order to stop the analysis when the boundary is no longer being sampled in the analysis volume. The specimen may then be reimaged so that additional analyses across the boundary may be performed. In analyses of both precipitates and interfaces, it is generally advisable to position the specimen to collect ions from the surrounding matrix so that solute profiles across the interface and into the matrix may be constructed and analyzed.

There are slight differences between the apparent position of the atom in the field ion image and the position where the ion strikes the single atom detector. The effect is most severe at the edges of atomic terraces, boundaries and at precipitate-matrix interfaces. This difference is referred to as the aiming error and should be taken into account when positioning the specimen for analysis. This effect arises since the positions where the ionization of the specimen atom and the image gas atom occurs are slightly different and the specimen atom acquires some energy (typically 60 eV) before passing through the critical ionization zone where the image gas is ionized. The former ionization process occurs at the specimen surface whereas, the latter occurs at a critical distance above the surface. In addition, the trajectory of the specimen atom may also be perturbed slightly by strong interactions with neighboring atoms as it leaves the surface, §5.5.3.

The feasibility of performing a composition measurement and its accuracy are dependent on the number of atoms available for analysis in a feature. An analysis typically contains between ~10^5 and ~2×10^6 atoms. Therefore, in a matrix analysis where there is a random distribution of solute, it is possible to detect the presence of an element with a concentration level as low as ~0.005 at. %, assuming that there are no peak overlap problems (§5.4.1). The precise detection limit is normally a function of the signal-to-noise ratio of the analysis (§5.4.4), the number of isotopes and different charge states exhibited by the element of interest, and the mass resolving power of the atom probe.

The number of atoms available for analysis in a small precipitate is considerably smaller than the number of atoms in the analyzed volume, as shown for a body centered cubic precipitate in Fig. 5.3. Therefore, the minimum solute concentration that can be detected in an individual precipitate is significantly larger. Similarly, the number of atoms available for analysis in a 0.2-nm-thick spherical shell at the precipitate-matrix interface is also shown in Fig. 5.3. The number of atoms available for analysis in Fig. 5.3 represent the theoretical limit; the actual number for a real system must take into account the efficiencies of the detection process. Also, if only a portion of the precipitate is within the analyzed volume, this number should be adjusted accordingly, as discussed in §5.6.

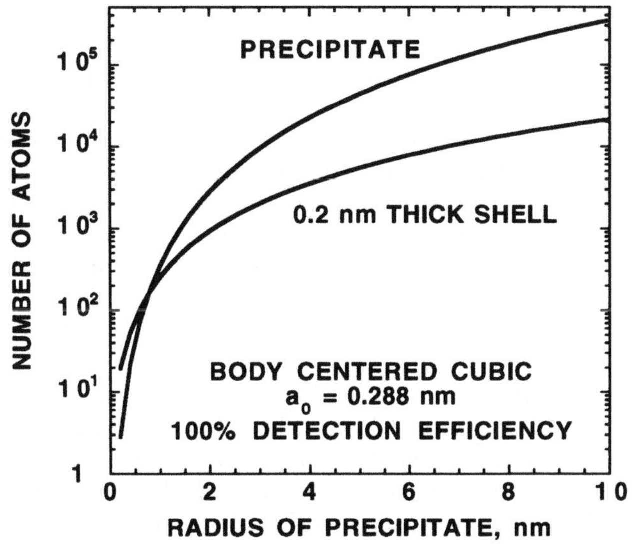

Fig. 5.3. Number of atoms available for analysis in a) a body centered cubic precipitate and the number of atoms in b) a 0.2-nm-thick shell at the precipitate-matrix interface.

5.3 Preferential Retention and Evaporation

In order to collect data that are free from artifacts, it is critical that the correct experimental conditions are used. The most important parameters in an atom probe experiment are the temperature of the specimen during analysis, the standing field or voltage and the pulse fraction applied to the specimen. If these parameters are not carefully chosen, incorrect compositional estimates may arise because elements with relatively small or large field evaporation potentials may be preferentially evaporated or retained, respectively.

From the image force model (§3.4.2), the field dependence, F, of the field evaporation rate constant for an n-fold charged ion, K_n at a fixed temperature, T, is given by [3]

$$\left(\frac{d\ln K_n}{d\ln F}\right)_T \approx \frac{Q_{0(n)}}{2k_B T}, \qquad 5.1$$

where, $Q_{0(n)}$, is the energy barrier to desorption under zero field conditions, and k_B is Boltzman constant. Since $Q_{0(n)} \gg k_B T$, this equation yields a power law with a large exponent. Therefore, there is a strong dependence of the field evaporation rate with field strength. For example, the evaporation rate of pure metals may increase by a factor of 10 for only a 1% change in the field. This behavior enables a relatively small increase in the field strength to change the specimen from a stable non-evaporating state to one where it is uniformly field evaporating. Because there is a practical limit to the amplitude of the high voltage pulse that may be generated, the standing voltage on the specimen must be set to the point at which no field evaporation is occurring and the pulse provides the necessary field increase to promote field evaporation.

A schematic representation of the effect of standing voltage amplitude, V_{dc}, and voltage pulse amplitude, V_{pulse}, on field evaporation in a multielement specimen is shown in Fig. 5.4. The basic premise for correct operation is that no ions should field evaporate at V_{dc} and that all atoms should have the same probability of field evaporation at $(V_{dc} + V_{pulse})$, as represented by case ① in Fig. 5.4. If the increase in field due to application of the pulse voltage is too low, case ②, the element that is more difficult to field evaporate (A) will be retained on the surface. As the important parameter is the increase in the field rather than the amplitude of the pulse, this parameter is usually expressed as a pulse fraction. If the standing field or voltage on the specimen is too high, case ③, some of the element (B) with the lower evaporation field will preferentially evaporate during the time between the pulses. Since there is a large time difference between the time during which the pulse is active (typically 10-20 ns) and the time between pulses (typically 0.66 ms at a pulse repetition rate of 1500 Hz), it is important to ensure that no evaporation occurs at the standing field. Since the field dependence of the evaporation rate becomes stronger at lower temperatures, lower pulse fractions are required at lower specimen temperatures. Therefore, the parameters that are appropriate at one temperature, T_2, may not be suitable at either higher, T_3, or lower, T_1, specimen temperatures and may result in preferential retention of element A, case ④, and preferential evaporation of element B, case⑤, respectively.

Faster pulse repetition rates, colder specimen temperatures, lower standing fields and higher pulse fractions tend to minimize the preferential evaporation and retention effects. However, too low a standing voltage can also lead to field corrosion and other forms of contamination. For a large variety of materials, a specimen temperature of 50 K, a pulse fraction of 20% and a pulse repetition rate of 1500 Hz provide a standard condition for analysis. It is customary to perform calibration experiments with a range of pulse

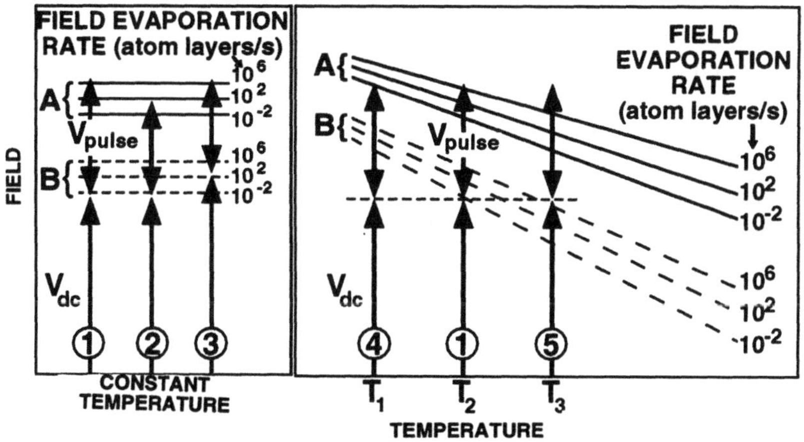

Fig. 5.4. Schematic diagram of the field evaporation rate as a function of specimen temperature for an A-B alloy. See text for details.

fractions and specimen temperatures to ensure that the conditions selected yield the correct overall composition of the alloy.

For some materials that exhibit a strong orientation dependence in the field evaporation rate, care should be exercised at the start of an atom probe experiment since the surface generated by field evaporation by increasing the standing voltage will not be the same as that produced by pulsed field evaporation. This effect leads to initially different field evaporation behaviors at different locations on the surface. Therefore, it is possible to increase the field on the specimen prematurely, which can lead to specimen failure.

5.4 Interpretation and Assignment of Ions

In order that the compositions determined in the three-dimensional atom probe be accurate, several factors must be taken into account in the assignment of the correct element to a given mass-to-charge ratio. These factors are discussed in this section.

The correct positioning of the specimen in the mass spectrometer is of critical importance for proper operation of the atom probe. The relationship of the specimen with the counter electrode defines the field distribution on the specimen. It is therefore essential that the specimen is placed precisely on the ion-optical axis of the instrument. This operation is either performed during loading the specimen into its holder prior to insertion into the atom probe or when the counter electrode is positioned in front of the specimen in the tomographic atom probe. Severely misaligned or bent specimens may not permit a portion of the specimen to be imaged or analyzed. The placement of the specimen along the ion-optical axis determines the distance traveled in the mass spectrometer and hence the calculated mass-to-charge ratio and the automatic assignment of elements to the peaks in the mass spectrum. A precise knowledge of the position of the specimen is also necessary for the accurate reconstruction of the three-dimensional data.

In order to identify the ions correctly, it is essential that their mass-to-charge ratios be accurately estimated from the flight times. The first step in this process is to ensure that the mass spectrometer and the estimated mass-to-charge ratios are accurately calibrated. It is normally only necessary to perform this calibration once for each instrument. However, if the specimen-to-counter electrode distance varies significantly, a new calibration is required. This calibration is required because it is difficult to accurately measure the flight distance in an energy-compensated atom probe and to determine the transmission times in the cables used to connect the time-to-digital converter, preamplifiers and discriminators. In addition, the coupling factor on the pulse voltage and the linearity of the analog-to-digital converters used to measure the voltages may not be reliably estimated.

The standard equation (Eqn. 1.3) for the mass-to-charge ratio, *m/n* in atomic mass units (u) of each ion may be expressed as

$$\frac{m}{n} = 1.9297 \times 10^{-4}(\alpha V_{\text{dc}} + \beta V_{\text{pulse}})\frac{(t-\tau)^2}{\delta^2} \quad . \qquad 5.2$$

In this equation, t is the flight time in nanoseconds measured by the time-to-digital converter, τ is the correction term that takes into account the differences in the start and stop signals due to transmission times in the electronics and cables. The parameter δ is the estimate of the flight distance in millimeters. The parameters α and β are corrections for the standing voltage, V_{dc} in volts and the coupling factor and correction for the pulse voltage V_{pulse}, respectively.

One method of estimating these parameters is to collect a series of time spectra (an accumulation of ions collected at a given flight time) over a range of standing voltages and pulse fractions from a simple material. The material chosen for this calibration experiment should contain several peaks at different mass-to-charge ratios from light and heavy elements and should not be prone to the formation of hydride species or other problems that complicate the assignment of the peaks to the correct mass-to-charge ratio. A simple robust material for this purpose is boron-doped Ni_3Al. This material exhibits peaks for B^{2+} at 5 and 5.5 u, B^+ at 10 and 11 u, Al^{3+}, Al^{2+} at 9 and 13.5 u and Ni^{2+} and Ni^+ peaks at 29, 30 30.5, 31, 32 and 58, 60, 61, 62 and

64 u, respectively. The flight times of ions in the time bins at the maxima of these peaks in the time spectra and their associated mass-to-charge ratios, in addition to the measured standing and pulse voltages, may be used to perform a multi-dimensional curve fit to estimate the parameters α, β, δ and τ.

Small differences in the position of the apex of the specimen along the optical axis of the instrument will give rise to a systematic change in the calculated mass-to-charge ratios due to the difference in the total flight distance. This change will be constant for a given specimen and the peak assignment routine should be able to accommodate this small change.

The assignment of the correct element to a given mass-to-charge ratio is normally performed with reference to the accumulated mass spectrum and tables of isotope abundances for the elements (Appendix G). The mass spectrum is simply a plot of the number of ions collected as a function of their mass-to-charge ratio, as shown in Fig. 5.5 for a nickel-molybdenum superalloy. It is common practice to plot mass spectra with a logarithmic abscissa as this emphasizes the minor and usually interesting peaks, as shown in Fig. 5.6. The mass range for each data point or bin size for the mass-to-charge ratio should be chosen to resolve neighboring peaks to their base. The closest separation between peaks, 1/6 u, occurs between the 3^+ and 2^+ species of different elements, such as $^{92}Mo^{3+}$ and $^{61}Ni^{2+}$. The number of cases of overlapping 4^+ and 3^+ peaks is extremely small and the elements involved are uncommon in metallurgical systems. Therefore, bin size of the mass spectrum is normally 0.01 to 0.02 u which yields peak to peak separations of ~16 and ~8 bins, respectively for peaks separated by 1/6 u.

Many elements field evaporate in more than one charge state, e.g., Al^{2+}, Al^{3+} and Al^+. Under normal field evaporation conditions, the highest charge state that is typically observed is $n = 4$ for Mo, Ta, W, Os, and occasionally for Zr, Nb, Hf, Re and Ir. In general, higher charge states are observed at lower specimen temperatures, at high evaporation rates, or during analyses of phases that have high evaporation fields. In some cases, overlaps between different charge states can be minimized by careful selection of specimen temperature and evaporation rate. However, preferential evaporation and retention effects (§5.3) should be the first consideration in the selection of these experimental parameters.

Some species associated with the specimen preparation process may also be detected. For example, H^+, H_2^+ and H_3^+ may be observed in the mass spectra of materials that have been electropolished and Ar^+ and Ga^+ may be observed in ion milled specimens. In addition, image gas species from the field ion imaging process such as Ne^+ (and the extremely rare Ne^{2+}) and He^+ may be observed particularly during the start of an analysis. These species are

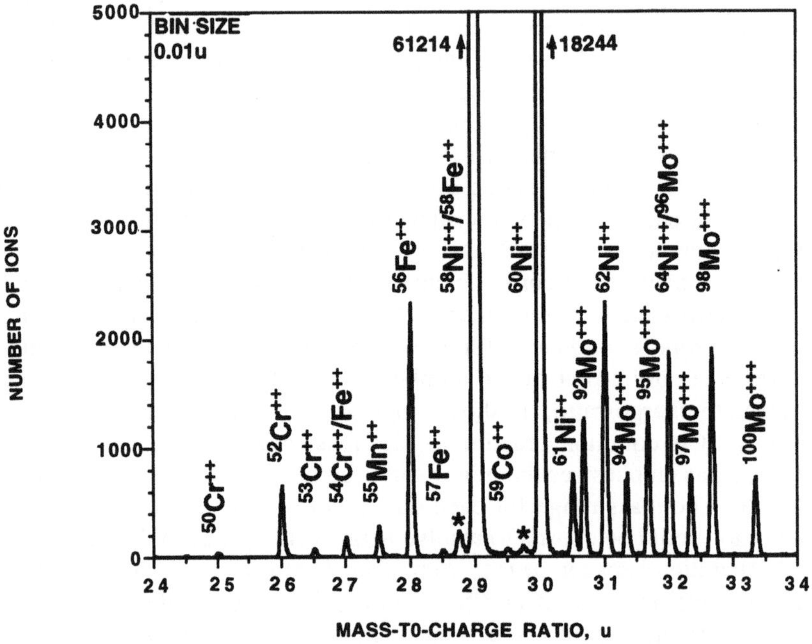

Fig. 5.5. Portion of a mass spectrum from a nickel-molybdenum superalloy obtained in an energy-compensated three-dimensional atom probe. Note the separation of the $^{92}Mo^{3+}$ and $^{61}Ni^{2+}$ peaks. Prepeaks indicated by * are evident to the low mass side of the main Ni^{++} peaks at 29 and 30 u due to the incorporation of a biased mesh in the single atom detector (see §5.4.5).

normally ignored in composition determinations since they are not characteristic of the material and do not overlap with any metallic species.

5.4.1 Isotope deconvolution

Not all the peaks in the mass spectrum can be unambiguously assigned to elements, as shown in Fig. 5.5, and therefore some peak deconvolution is normally required. For example, several elements may share the same isobar, e.g. $^{40}Ar/^{40}Ca$, $^{50}Ti/^{50}V/^{50}Cr$, $^{54}Fe/^{54}Cr$, $^{58}Fe/^{58}Ni$, $^{64}Ni/^{64}Zn$, $^{96}Zr/^{96}Mo/^{96}Ru$, etc. In addition, due to the multiple charge states that are commonly observed, some additional overlaps are possible. Some common examples are $^{14}N^{+}/^{28}Si^{2+}$ at 14 u, $^{60}Ni^{2+}/^{90}Zr^{3+}$ at 30 u, $^{62}Ni^{2+}/^{93}Nb^{3+}$ at 31 u and $^{64}Ni^{2+}/^{96}Mo^{3+}$ at 32 u. In most cases, the number of ions of each species in these overlapping peaks can be estimated from the number of ions collected in the other isotopes of the same charge state and the tables of isotope abundances of the elements. For example in the case of Ni^{2+}/Mo^{3+} overlap shown in Fig. 5.5, nickel has 5 isotopes at 58, 60, 61, 62 and 64 u and molybdenum has

7 isotopes at 92, 94, 95, 96, 97, 98 and 100 u. Therefore, there is an overlap between the $^{64}Ni^{2+}$ and $^{96}Mo^{3+}$ isobars at 32 u. The number of ions in the $^{64}Ni^{2+}$ peak, n_i, may be estimated from the total number of ions, n_o, in the peaks of the other nickel isotopes ($^{58}Ni^{2+}$, $^{60}Ni^{2+}$, $^{61}Ni^{2+}$ and $^{62}Ni^{2+}$) since the relative abundances of all the isotopes, f_i, are known (Appendix G), i.e.,

$$n_i = \frac{f_i}{1-f_i} n_o . \qquad 5.3$$

Similarly, the number of ions in the $^{96}Mo^{3+}$ peak can be estimated from the number of ions in the peaks of the other molybdenum isotopes ($^{92}Mo^{3+}$, $^{94}Mo^{3+}$, $^{95}Mo^{3+}$, $^{97}Mo^{3+}$, $^{98}Mo^{3+}$ and $^{100}Mo^{3+}$). The total of the numbers of ions estimated for the $^{64}Ni^{2+}$ and $^{96}Mo^{3+}$ isobars should be within the statistical error of the number of ions detected in the peak at 32 u (§6.1.4). Isotope overlaps may be eliminated through the use of isotopically tailored alloys. However, this approach is seldom used due to cost considerations.

5.4.2 Molecular and complex ion formation

Some of the ions that are field evaporated contain more than one atom. These ions are termed molecular ions. Molecular ion formation is particularly common in the case of carbon containing materials as carbon forms several molecular ions, C_3^{2+}, C_3^{+}, C_2^{+}, C_4^{2+} and C_2^{2+}. Molecular ion formation can be much more prevalent in semiconducting materials, especially when a pulsed laser is used to field evaporate the specimen. Some elements also form complex molecular ions, MN^{n+}, with other elements in the alloy or with adsorbed species on the surface of the specimen. These species are predominantly observed when there is a locally high concentration of the elements present such as in carbides, nitrides, oxides, etc. These species may also be formed through interactions with the residual gases in systems with poor vacuum conditions. Some relatively common examples are VC^{2+}, NbC^{2+}, VN^{2+}, TiO^{2+}, CrN^{2+}, MoN^{2+} and MoN^{3+}.

Many elements are also prone to the formation of hydride species, i.e., MH^{+}, MH_2^{+}, etc. Hydrides are particularly common in instruments with vacuums in the 10^{-9} and 10^{-10} mbar ranges or when hydrogen has been used as an image gas. Beryllium is known to form a $BeH.H_2^{+}$ species at 12 u. Helides, neides and argides have been observed occasionally when analyses have been performed in the presence of image gas in a classic atom probe. However, they are not normally observed in analyses performed under ultrahigh vacuum conditions. Additional overlaps in the mass spectrum may arise due to the formation of hydrides or molecular ions, e.g. $^{27}Al^{1}H^{2+}/^{28}Si^{2+}$, $^{12}C_2^{+}/^{48}Ti^{2+}$ and $^{51}V^{14}N^{2+}/^{65}Cu^{2+}$ species.

5.4.3 Ion pile up

When more than one ion strikes the single atom detector at precisely the same time, only one event is recorded by the timing system. This phenomenon is referred to as ion pile-up and results in a loss of ions. This problem is most severe in three-dimensional instruments that have large area anodes such as the position-sensitive atom probe and the original version of the optical atom probe. In instruments with secondary detectors, the composition should be determined from a mass spectrum accumulated from the primary detector that has been corrected for the multiple hits detected on the secondary detector.

The results of ion pile-up are more severe in materials that contain a predominant isotope because there is a high probability that two or more ions of that element will field evaporate on the same field evaporation pulse. The effect produces a reduction in the concentration of that element and a complementary increase in the concentration of the other elements.

Several statistical methods have been developed to correct the composition determined with the classic atom probe [4-6]. In the method developed by Menand *et al.* [6], the experimentally observed ratio of the number of double ion events to the total number of detected ions, n_d is used. In a mass spectrum containing I peaks each containing a fraction f_i of the ions, the ratio of double events lost, R_d, to those that are observed is given by

$$R_d = \frac{\sum_{i=1}^{I} f_i^2}{1 - \sum_{i=1}^{I} f_i^2}, \qquad 5.4$$

where the numerator and denominator are the proportion of double events involving similar type species and the proportion involving dissimilar types, respectively. The corrected solute concentration, c^*, may be estimated from the measured solute concentration, c, by

$$c^* = \frac{c}{1 + R_d n_d} + \frac{R_{d1} n_d}{1 + R_{d1} n_d}, \qquad 5.5$$

where the ratio R_{d1} is equivalent to Eqn. 5.4, except that the numerator is summed over all the peaks of that species. For low concentration species, the second term becomes negligible and may be ignored. These methods could also be applied to the data collected on individual anodes of a multianode detector if multiple channels or multi-hit channels of time-to-digital converters are assigned to each anode. However, these schemes have not been generally applied to the data from three-dimensional atom probes.

5.4.4 Noise and background substraction

Background noise limits the minimum detection level in an analysis. Compared to many other types of mass spectrometers and other techniques, the spectra produced in the atom probe are relatively free from background noise. Some of these sources can be minimized by operating at the lowest possible pressures and maintaining the highest evaporation rate without introducing other systematic errors, such as ion pile-up or preferential evaporation.

Although three-dimensional atom probe analyses are performed under UHV conditions, there is still some residual gas in the system. The time required to form a monolayer coverage on the surface of a specimen from the gas phase is 10^4 s at 10^{-10} mbar. The processes that occur during the formation of a field ion image significantly reduce this time. Therefore during an extended three-dimensional atom probe experiment that lasts several hours, many atoms from the gas phase that approach the apex of the specimen will be polarized, then attracted to the specimen, ionized and repelled towards the single atom detector. Some of these ions, such as residual image gas atoms, will be field evaporated during the high voltage pulse and appear at specific mass-to-charge ratios. These ions can be simply identified in the mass spectrum and ignored in compositional analyses. The standing voltage on the specimen will also ionize some atoms and because there is no correlation with the field evaporation pulse, they will appear as random noise in the time spectrum. It is possible to filter out these ions in a Poschenrieder type of energy-compensating lens but this filtering cannot be easily accomplished in a wide angle reflectron lens or a non energy-compensated atom probe. In addition to the ions originating in the gas phase, ions will also be generated from gas absorbed on the shank of the specimen that diffuses along the shank of the needle to the apex region.

Another source of random noise is the random noise (dark current) produced in the single atom detector. According to the data provided by the manufacturers of microchannel plates, typical noise levels are much less than 1 count $s^{-1}cm^{-2}$. This value gives a rate of less than ~100 counts per second for a typical 10 by 10 cm square single atom detector. However, the gating system used in the timing system ensures that the noise is only observed during the active period after each high voltage pulse. A typical 30 μs timing window would produce on average ~0.003 noise events per pulse. In practice, total noise levels of ~0.0002 events per pulse are typically measured in three dimensional atoms probes with 10 by 10 cm square single atom detectors and base pressures of less than ~2 x 10^{-11} mbar. This significantly lower experimental noise level indicates that the dark noise in the single atom detector is approximately an order of magnitude lower for this type of application than that reported by the manufacturers.

It is also possible that some ions emitted from the specimen will hit the walls of the vessel or other internal components before striking the single atom detector. Some of these ions may originate from outside the area of analysis. The flight times of these ions will be longer than expected and will lead to a high mass tail in the peaks in the mass spectra. The use of collimators in the mass spectrometer minimizes these effects.

Single atom detectors may generate signals due to exposure to electrons, x-rays and visible light; therefore, direct exposure to these sources of radiation should be eliminated. Electrical noise can also be generated at the detector and associated components from the high voltage pulse, electrical discharges in the system, and other sources such as neighboring equipment that contain heavy duty relays or motors. The threshold setting on the discriminator connected to the preamplifier on the single atom detector (that defines the minimum amplitude input pulse that will generate an output pulse) and the length of initial veto period after the high voltage pulse are important parameters in minimizing the electronic noise in the timing system.

In compositional analyses, background noise subtraction may be undertaken by examining portions of the mass spectrum, which do not contain any mass peaks. The number of events in these mass bins are then used to construct the background noise curve over the entire mass range so that the number of noise events in the mass bins that contain genuine mass peaks can be subtracted from the measured number of ions. A typical mass spectrum with a superimposed background noise curve is shown in Fig. 5.6. At low mass-to-charge ratios, the background noise is higher than at higher mass-to-charge ratios due to the larger number of timing bins per mass channel, as discussed in §4.3.4. The background noise subtraction will affect the smaller amplitude mass peaks more than the major peaks since the signal-to-background noise ratio is significantly smaller.

5.4.5. Prepeaks

In single atom detectors that incorporate biased high transparency meshes to increase the detection efficiency (§4.3.2), a small amplitude peak is evident to the low mass side of major peaks in the mass spectra. An example obtained from an energy-compensated three-dimensional atom probe is shown in Fig. 5.5. This prepeak arises from ions impacting the mesh and generating electrons which are then accelerated rapidly towards and detected on the single atom detector at a slightly earlier time than the heavier ion. The shift in the measured mass-to-charge ratio, Δm, for an ion hitting the mesh is given by [7]

$$\Delta m = \frac{m}{n}\left[1-\left(\frac{\delta - d_m}{\delta}\right)^2\right], \qquad 5.6$$

where d_m is the distance between the mesh and the detector and δ is the specimen-to-detector distance. This equation assumes that the time taken for the electron to traverse the distance from the mesh to the front surface of the detector is negligible. The difference in the mass-to-charge ratio for an ion hitting a mesh placed at a typical distance of 2 mm in front of the detector, is shown in Fig. 5.7 for an atom probe with typical flight distance of δ = 0.62 m. The position of the prepeaks decreases proportionally with the mass-to-charge ratio of the ion and is independent of the applied tip and pulse voltages. Experiments have revealed that the prepeak contains ~0.3% of ions in the main peak for a 98% transparent mesh. Therefore, these prepeaks eliminate a proportion (15%) of the loss due to ions striking the mesh and increase the effective detection efficiency. However, these prepeaks must be properly identified and not misinterpreted as arising from a distinct element. The intensity in the prepeaks can be reduced by coating the mesh with a material having a low secondary electron yield, such as carbon.

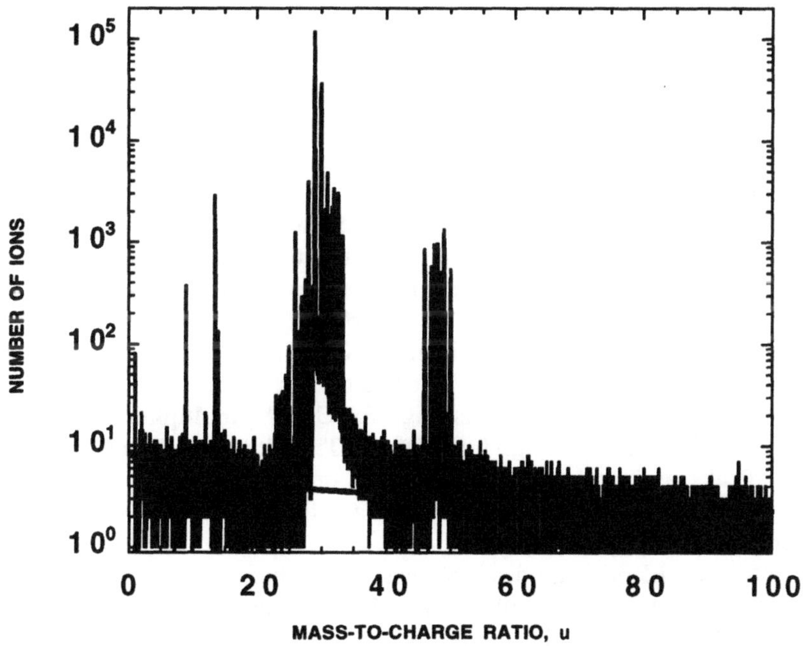

Fig. 5.6. A mass spectrum from a nickel-molybdenum superalloy with a background noise fit curve superimposed. Note the use of a logarithmic scale. This spectrum contains a total of 1,995,322 ions.

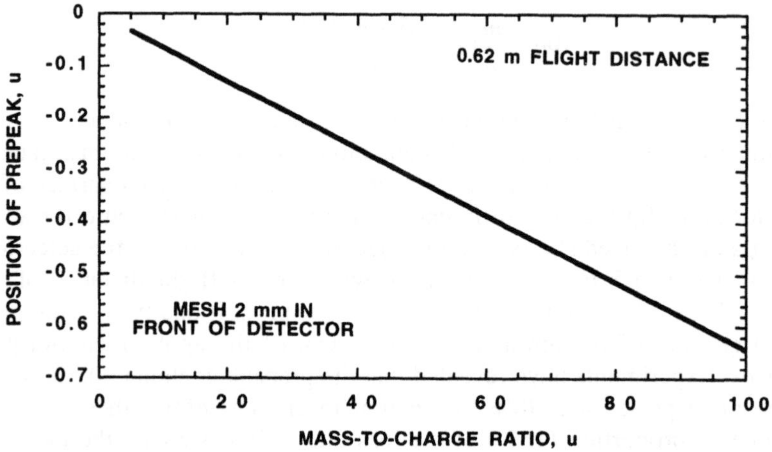

Fig. 5.7. The difference in the measured mass-to-charge ratio for an ion hitting a mesh placed at a typical distance of 2 mm in front of the single atom detector.

5.4.6 Windowing system

The identification of the ions from their mass-to-charge ratio is often performed with the use of a window or range file. This file contains a list of the material related parameters including the names of the elements, mass ranges and multiplying factors for the molecular ions. Additional information, such as the color used to represent each element, plot symbols, and the atomic densities of the elements may also be included in the range file. An example of a range file for a nickel aluminide containing boron and niobium additions is shown in Fig. 5.8. As in this example, it is common to group contiguous and non-overlapping isotopes of one element into a single mass window in order to reduce the number of mass windows. Some of the ranges are assigned to peaks containing more than one element due to isotope overlaps to enable peak deconvolution to be performed. The multipliers can take into account molecular ions (e.g., for C_2^+, the multiplier = 2) and also complex molecular ions (e.g., for TiO^+, the Ti multiplier = 1 and O multiplier = 1). Peaks assigned to image gas ions or other spurious peaks can be assigned multipliers of zero to eliminate them from the compositional determinations.

Automatic identification of the mass peaks has also been accomplished with the use of databases of normal charge states observed and tables of natural abundance of the elements. The identification is performed by accumulating the mass spectrum, performing a background noise subtraction as described in §5.4.4, aligning the mass spectrum to the correct masses,

```
6 9                          : Number of elements and mass windows
Boron 130.                   : Element name, atomic density (atoms nm-3)
B    1.00 1.00 1.00          : Element symbol, RGB colors
Aluminum 60.2
Al   0.00 1.00 0.00          : RGB color (in red, green, blue intensities)
Nickel  91.4
Ni   1.00 0.00 0.00
31                           : Nb3+ and 62Ni++ common isobar at 31 u
31   0.00 1.00 1.00
Niobium  55.6
Nb   1.00 1.00 0.00
Neon  0
Ne   0.10 0.10 0.10
----------------B  Al Ni 31 Nb Ne    : comment line
B    4.90    5.65 1  0  0  0  0  0   : symbol, mass start & end
B    9.90   11.15 1  0  0  0  0  0   : and multipliers
a    8.90    9.15 0  1  0  0  0  0
a   13.40   13.65 0  1  0  0  0  0
a   26.90   27.15 0  1  0  0  0  0
•   28.90   30.89 0  0  1  0  0  0   : 29, 30 and 30.5 u
•   31.16   32.15 0  0  1  0  0  0
3   30.90   31.15 0  0  0  1  0  0   : common peak at 31 u
N   46.40   46.65 0  0  0  0  1  0
-   19.90   22.15 0  0  0  0  0  0   : multipliers are 0 for neon
```

Fig. 5.8. Example of a range file for a nickel aluminide containing boron and niobium additions.

identifying the mass ranges of all the peaks in the mass spectrum, assigning possible elements to each peak based on the databases of normal charge states observed and tables of natural abundance of the elements, and eliminating possible elements based on the absence of other isotopes of that element. The peak identification process may also be used to automatically generate range files.

5.5 Reconstruction of Atom Positions

In this section, the procedures used to estimate the positions of the atoms in the specimen from the parameters measured from the single atom detector are discussed.

The x and y directions are defined to be orthogonal to the specimen axis and the z direction is defined to be along the specimen axis. Instrument dependent procedures to convert the raw measurements, such as charge on the electrodes or pixels in the CCD cameras, into x and y coordinates are discussed in §4.4. These x and y coordinates are translated so that the origin of the system is projected through the center of the single atom detector. In addition, the relative X_a and Y_a coordinates are scaled to be in true distance units, i.e.,

$$X_d = X_a d_x - \frac{d_x}{2} \quad \textit{and} \quad Y_d = Y_a d_y - \frac{d_y}{2}, \qquad 5.7$$

where d_x and d_y are the physical extents of the active area of the single atom detector in the x and y directions, respectively.

5.5.1 Basic method

The x and y coordinates of the position of the ion in the specimen for an analysis along the specimen axis may be expressed [8,9] as

$$x = \frac{X_d}{\eta} \quad \text{and} \quad y = \frac{Y_d}{\eta}, \qquad 5.8$$

where η is the magnification of the image on the detector and is given by

$$\eta = \frac{\delta}{\xi r_t}, \qquad 5.9$$

where the parameter ξ is a projection parameter known as the image compression factor, δ is the estimated specimen to single atom detector distance and r_t is the radius of the specimen. Since the field of view on the single atom detector is limited, the image compression factor is normally determined from a field ion image on the field ion detector, as discussed in §3.10.3, and assumed to be the same for the single atom detector.

As the specimen is field evaporated, its radius normally increases and hence the magnification decreases. Since it is impractical to measure the radius of the specimen after each ion is field evaporated, it is estimated from the evaporation voltage, V, applied to the specimen with the following relationship (Eqn. 1.1)

$$r_t = \frac{V}{k F}, \qquad 5.10$$

where F is the evaporation field at the apex of the specimen and k is the geometrical field factor. This relationship is, to a first approximation, a linear relationship with the voltage. Therefore, if the radius of the specimen is independently determined for a particular evaporation voltage either by examining the specimen in the transmission electron microscope or by counting the number of rings between known poles in a field ion micrograph, as discussed in §2.7 and §3.10.1, respectively, the parameter kF can be estimated. The radius at any other voltage can then be extrapolated from Eqn. 5.10. If the radius of the specimen at a particular evaporation voltage cannot be independently determined due to specimen failure, etc., estimates of the evaporation field, F, and geometrical field factor k have to be made. Some predictions of the evaporation field for the elements based on the image hump model [10] are given in Appendix C. These values may also be used for dilute alloys. However, calibration experiments in which the radius of the specimen is measured for a particular evaporation voltage are required for concentrated alloys and other phases. The geometrical field factor depends primarily on the taper angle of the specimen. The value of the geometrical field factor increases from ~2 for a parallel sided specimen up to ~8 for a specimen with a large taper angle.

The z position is determined from the order in which the ions strike the single atom detector. For each ion impacting a square detector, the position of the ion in the z direction is incremented by an amount z_i, which is given by

$$z_i = \frac{\Omega_i}{\zeta A_d} = \frac{\Omega_i}{V_i^2}\left[\frac{(\delta\, kF)^2}{\zeta\, d_x d_y\, \xi^2}\right], \qquad 5.11$$

where Ω_i is the atomic volume of the ith ion in the phase under analysis, V_i is the effective voltage applied to the specimen to field evaporate the ion, A_d is the effective area and ζ is the detection efficiency of the single atom detector (§5.6). For a circular single atom detector, $d_x d_y$ is replaced by $\pi d_x^2/2$, where d_x is the diameter of the detector. The factor enclosed in braces in Eqn. 5.11 is a constant for a given analysis since all the parameters are either instrument or material dependent. In order not to inflate the resulting volume, only atoms that originate within the specimen should be used in this calculation. Other atoms such as image gas atoms, implanted atoms that arise from the specimen preparation processes and background noise should be excluded. One method that may be used to eliminate these ions is to set Ω_i to zero for all ions not having appropriate mass-to-charge ratios. In addition, applicable atomic volumes should be used for interstitial ions.

In the analysis of single phase materials, Ω_i is usually replaced with the average atomic volume of the material, $\overline{\Omega}$, to simplify the calculation. The atomic volume may be calculated from the volume, v_u, and number of atoms, n_u, in the unit cell of the structure, i.e., $\Omega = v_u/n_u$. The formulae for estimating the volumes of the unit cells of the different crystal structures from their unit cell parameters are listed in Appendix A. For example in the $L1_2$-ordered Ni_3Al phase, there are 4 atoms per unit cell and the lattice parameter, a_o, of the cubic unit cell is 0.356 nm. Therefore, the atomic volume is $0.356^3/4=$ $0.011\ nm^3$. The atomic volume may also be determined from the inverse of the atomic density, γ, i.e., $\Omega = 1/\gamma$. The atomic densities of the elements are given in Appendix F.

In multiphase materials where the phases have different atomic volumes, it is possible to adjust the atomic volumes for the different phases by the use of multiple passes through the ion-by-ion data. In the first pass, an average value of the atomic volume is used to approximately position the ions. The local distribution of the ions is then examined to identify the ions that belong to each phase, as discussed in §6.1. In the second pass, ion-by-ion data is read again and z positions are determined with the atomic volumes of the different phases. Therefore, the z position of the nth atom detected is the sum of the z increments of all the atoms detected up to that atom, i.e.,

$$z = \sum_{i=1}^{n-1} z_i . \tag{5.12}$$

A small correction, z', has to be applied to the z position due to the curvature of the specimen surface, as shown in Fig. 5.9, such that

$$z = z + z' = z + \left(r_t - \sqrt{r_t^2 - (x^2 + y^2)}\right), \tag{5.13}$$

assuming that the specimen has a hemispherical end cap. This correction is referred to as the radius or "bent plane" correction since without this correction the atomic planes in the reconstruction are curved rather than planar, as shown in Fig. 5.10.

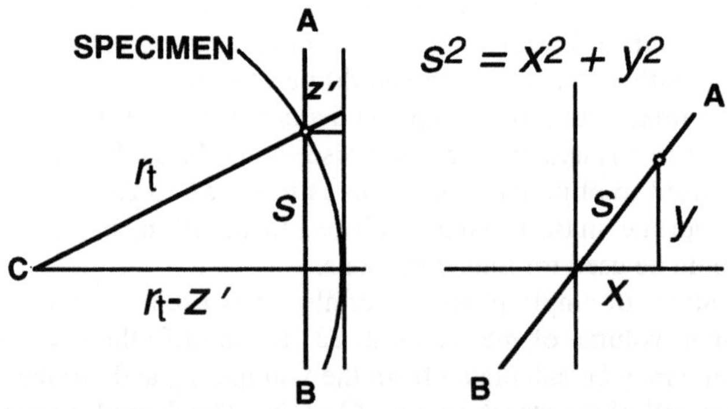

Fig. 5.9. Correction in the atom position to take account of the radius of curvature of the specimen. The AB plane shown on the right is orthogonal to the axis of the specimen and the section on the left has been rotated about the specimen axis such that line AB is vertical. Note that the correction has been exaggerated for clarity.

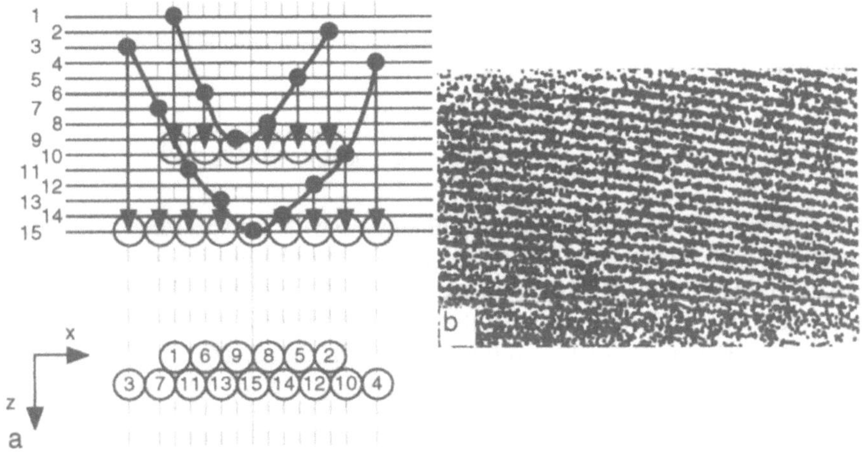

Fig. 5.10. a) The correction to take account of the curvature of the specimen to prevent "bent" planes. Examples of atom map reconstructions of planes in an aluminum specimen, b) before correction and c) after correction.

The basic method described above assumes that the analysis is performed with the specimen axis coaxial with the center of the single atom detector. However, many analyses are performed at off axis positions with the specimen rotated so that the central axis of the analyzed volume is inclined at angles of ψ_1 and ψ_2 to the specimen axis. Therefore, these rotations must be taken into account in the reconstruction [9]. The x and y positions may be given for an analysis in the tomographic atom probe with rotations of ψ_1 and ψ_{2T} by

$$x = x_p \sqrt{\frac{z'(z'-2r)}{x_p^2 + y_p^2}} \quad \text{and} \quad y = y_p \sqrt{\frac{z'(z'-2r)}{x_p^2 + y_p^2}}, \qquad 5.14$$

where

$$x_p = \left(\frac{X_d}{\eta_\psi} \cos\psi_{2T} + \frac{Y_d}{\eta_\psi} \sin\psi_{2T} \right) + \xi' \, r \sin\psi_1,$$
$$y_p = -\frac{X_d}{\eta_\psi} \sin\psi_{2T} + \frac{Y_d}{\eta_\psi} \cos\psi_{2T}, \qquad 5.15$$

$\xi' = \xi - 1$ and $\eta_\psi = \delta/(\xi'\cos\psi_1 + 1)r_t$. The z position increment along the direction of analysis may be determined with a modified version of Eqn. 5.11 such that

$$z_i = \frac{\Omega_i}{V_i^2}\frac{(\delta\, kF)^2}{\zeta\, d_x d_y}\frac{(1+\xi'\cos\theta_1)}{(1+\xi'^2+2\xi'\cos\theta_1)^{3/2}}\frac{1}{\cos(\theta_1-\phi)}. \qquad 5.16$$

The direction of analysis makes a small angle ϕ with the specimen axis. For a specimen with a taper angle of α, the angle ϕ may be determined [1,9] from $\tan\phi = (\sin\theta_1 \sin\alpha/2)/(1-\cos\theta_1 \sin\alpha/2)$ where θ_1 is the angle between the specimen axis and the normal to the hemispherical specimen surface. The angle θ_1 may be evaluated from the relationship $\sin(\theta_1 - \psi_1) = \xi' \sin\psi_1$. The z position is then given with the combination of Eqns. 5.12, 5.13 and 5.16.

5.5.2. Stereographic projection method

An alternative approach that involves projections may be used to determine the coordinates of the atom in the specimen. In the case of the stereographic projection, shown in Fig. 5.11, each atom on the surface of the specimen, I, is projected from point P onto the plane of the detector. By construction of a plane (S′QP) through the impact position, the center of the detector and the projection point, a relationship between the positions of the atom in the specimen and its impact on the detector may be determined such that

$$\frac{p}{P} = \frac{r_t(1+\cos\theta)}{2r_t+\delta} = \frac{2r_t}{2r_t+\delta}\cos^2\tfrac{\theta}{2} = \frac{2r_t}{2r_t+\delta}\left(\frac{1}{1+\tan^2\frac{\theta}{2}}\right), \qquad 5.17$$

$$\text{where} \quad \tan\tfrac{\theta}{2} = \frac{P}{2r_t+\delta}. \qquad 5.18$$

$$\therefore \quad \frac{p}{P} = \frac{2r_t(2r_t+\delta)}{P^2+(2r_t+\delta)^2}. \qquad 5.19$$

$$\text{Since} \quad \frac{p}{P} = \frac{x}{X_d} = \frac{y}{Y_d} = \frac{1}{\eta} \quad \text{and} \quad P^2 = X_d^2 + Y_d^2, \qquad 5.20$$

the x and y positions may be expressed as

$$x = \frac{2r_t(2r_t+\delta)}{X_d^2+Y_d^2+(2r_t+\delta)^2}X_d \quad \text{and} \quad y = \frac{2r_t(2r_t+\delta)}{X_d^2+Y_d^2+(2r_t+\delta)^2}Y_d \quad , \qquad 5.21$$

where r_t is estimated from Eqn. 5.10.

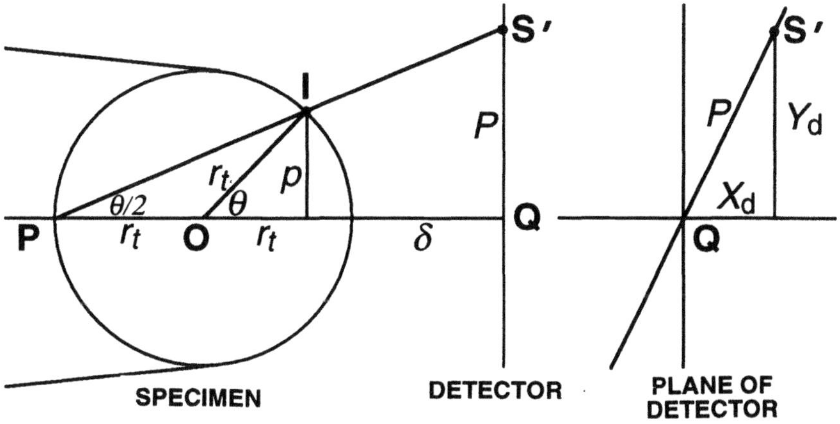

Fig. 5.11. Correction for the rotation of the specimen from the optical axis of the instrument.

The z position may be estimated with an approach similar to that in the basic method, that is Eqns. 5.11, 5.12 and 5.13, although with a slightly modified equation for the z increment,

$$z_i = \frac{\Omega_i}{\zeta A_d} = \frac{\Omega_i}{\zeta} \frac{\eta^2}{d_x d_y} = \frac{\Omega_i}{V_i^2} \frac{((\delta + 2r_t)kF)^2}{\zeta d_x d_y (1+\cos\theta)^2}. \qquad 5.22$$

This method provides similar results to the method described in the previous section.

In the general case shown schematically in Fig. 5. 12, the analysis is undertaken with the specimen inclined at angles ψ_1 and ψ_2 to the specimen axes VOW and UOT, respectively. By applying a stereographic projection, the position of an atom on the specimen surface may be determined with a two step procedure [11]. In the first step, the position of the projected atom is determined in terms of detector coordinates. The coordinates x_s and y_s of the atom on the specimen surface at point I are given by

$$x_s = x_o + A \quad \text{and} \quad y_s = y_o + B, \qquad 5.23$$

where $A = r_t \sin\psi_1 \cos\psi_2$ and $B = r_t \sin\psi_2$.

$$\tan\theta_y = \frac{y_o}{z_s + G} = \frac{y_s - A}{z_s + G} = \frac{Y_d}{\delta + r_t + G} = C \qquad 5.24$$

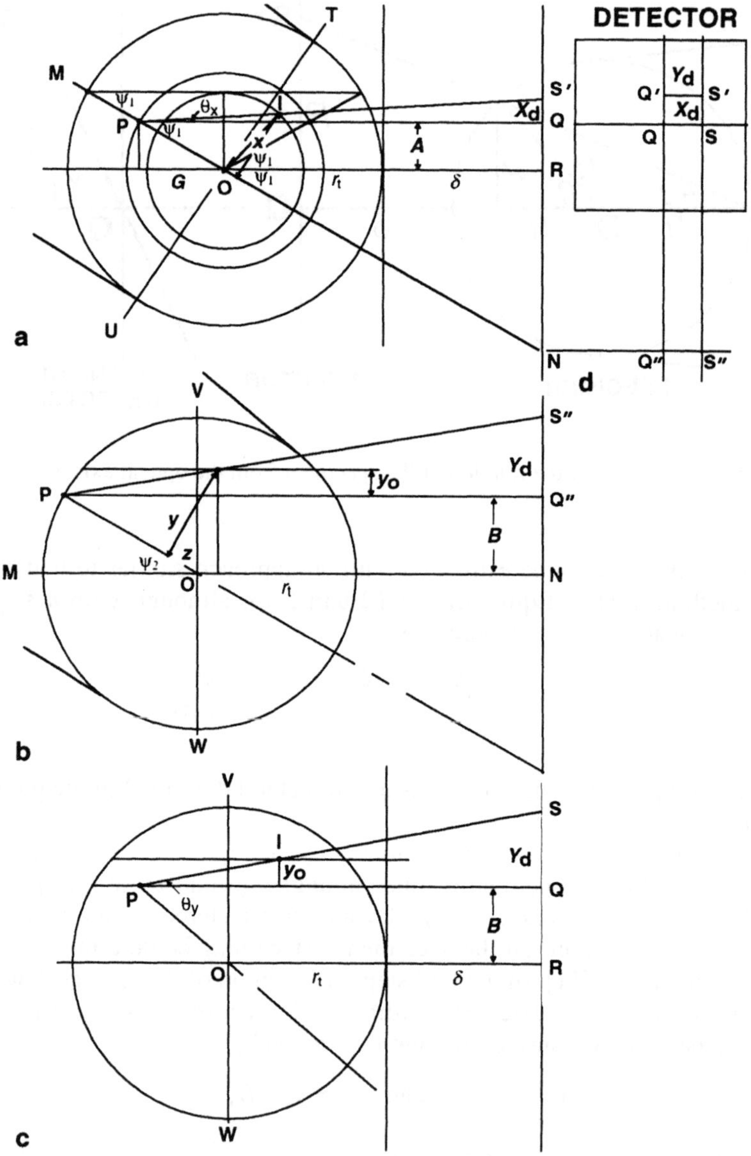

Fig. 5.12. Gnominc projections of the geometrical model used for the stereographic projection of ions from the specimen to the single atom detector in which the specimen is inclined at angles ψ_1 and ψ_2 to the specimen axes VOW and UOT, respectively. The specimen-to-detector distance, δ, is significantly larger than indicated.

$$\tan\theta_x = \frac{x_o}{z_s + G} = \frac{x_s - A}{z_s + G} = \frac{X_d}{\delta + r_t + G} = D, \tag{5.25}$$

where $G = r_t \cos\psi_1 \cos\psi_2$. The specimen coordinates may be related to the impact coordinates, X_d and Y_d, of the ion on the detector at point S′ by

$$\frac{x_o}{X_d} = \frac{y_o}{Y_d}. \tag{5.26}$$

$$\therefore \quad \frac{y_o}{x_o} = \frac{Y_d}{Y_d} = \frac{y_s - B}{x_s - A} = E, \tag{5.27}$$

$$y_s = Ex_s - (EA - B). \tag{5.28}$$

As the atom is on the surface of a hemispherical cap,

$$z_s^2 = r_t^2 - x_s^2 - y_s^2 = r_t^2 - x_s^2 - [Ex_s - (EA - B)]^2. \tag{5.29}$$

By substituting $z_s = (x_s - (A + DG))/D$ from Eqn. 5.25 into Eqn. 5.29 and rearranging the terms, the following quadratic equation is obtained

$$\begin{aligned} &x_s^2[1 + D^2(1 + E^2)] + 2x_s[ED^2(B - EA) - (A + DG)] \\ &+ [(A + DG)^2 + D^2((EA - B)^2 - r_t^2)] = 0. \end{aligned} \tag{5.30}$$

The solutions of this quadratic equation are

$$\begin{aligned} x_s &= \frac{-b + \sqrt{b^2 - 4ac}}{2a}; \quad X_d > 0 \\ x_s &= \frac{-b - \sqrt{b^2 - 4ac}}{2a}; \quad X_d < 0, \end{aligned} \tag{5.31}$$

where

$$\begin{aligned} &a = [1 + D^2(1 + E^2)], \\ &b = 2[D^2E(B - EA) - (A + DG)] \text{ and} \\ &c = (A + DG)^2 + D^2\left[(EA - B)^2 - r_t^2\right\}. \end{aligned} \tag{5.32}$$

Therefore in these cases, the position of the atom is given in detector coordinates by Eqns. 5.31, 5.28 and 5.29.

If $x_a = 0$ then $x_s = A$ and y_s is determined from the following relationships.

$$z_s^2 = r_t^2 - A^2 - y_s^2, \qquad 5.33$$

$$z_s = \frac{1}{C}(y_s - B) + G, \qquad 5.34$$

$$y_s^2(1 + C^2) - 2(B + CG)^2 y_s + (B + CG)^2 + C^2(A^2 - r_t^2) = 0, \qquad 5.35$$

$$\begin{aligned} y_s &= \frac{-b + \sqrt{b^2 - 4ac}}{2a}; \quad Y_d > 0, \\ y_s &= \frac{-b - \sqrt{b^2 - 4ac}}{2a}; \quad Y_d < 0, \end{aligned} \qquad 5.36$$

where

$$\begin{aligned} a &= 1 + C^2, \\ b &= -2(B + CG) \quad \text{and} \\ c &= (B + CG)^2 + C^2(A^2 - r_t^2). \end{aligned} \qquad 5.37$$

In this case, the position of the atom is given in detector coordinates by $x_s = A$ and by Eqns. 5.37 and 5.34.

The second step of this process is to convert the position of the atom from detector coordinates into specimen coordinates with the following transformation matrix

$$(x, y, z) = (x_s, y_s, z_s)\begin{pmatrix} \cos\psi_1 & -\sin\psi_1 \sin\psi_2 & -\sin\psi_1 \cos\psi_2 \\ 0 & \cos\psi_2 & -\sin\psi_2 \\ \sin\psi_1 & -\sin\psi_2 \cos\psi_1 & -\cos\psi_2 \cos\psi_1 \end{pmatrix}. \qquad 5.38$$

The z increment along the specimen axis is determined in the same manner as described above.

5.5.3 Ion trajectories

The methods described in the previous sections assume that the impact position on the single atom detector has a linear relationship with its position in the specimen. However, the trajectory of the ion between the specimen and the single atom detector is curved because it follows the field lines. In addition, both field evaporation images (§3.2.9) and the distributions of ions on the single atom detector show that local deviations occur in the trajectories as the ion leaves the specimen. These deviations occur due to the influence of the field on the neighboring atoms on the surface. It is also possible that the ion may move slightly along the surface prior to field evaporation, as shown in Fig. 5.13 [12]. In theory, these effects could be modeled if the positions of all the atoms in the specimen are accurately known. Although a full treatment of this type of trajectory method is not currently practical, some modeling has been performed. In these simulations, the trajectories of the field evaporated ions are disturbed by small repulsive forces from the ions' nearest neighbors on the specimen surface [13]. The magnitudes of the repulsive forces were modeled as a function of the nearest neighbor distances. These simulations exhibit many of the lineal features such as the bright and dark lines, the double lines, and the prominent ring patterns observed in the field desorption images. Another type of computer simulation that models the trajectory of ions has recently been developed. In the initial implementation of this model, the apex region of the specimen was represented by a series of circular slabs whose diameters initially conformed to that of a hemispherical cap. The trajectories of the ions near the center of a pole were found to have a linear relationship of the form $L_p = g\theta + L_o$ between the position of impact on the detector, L_p, and the angle from the center of the pole, θ, where g is a constant [14]. This linear relationship established that there was an offset, L_o, in the image of the pole from the center of the detector. This result was applied to three-dimensional atom probe data obtained near a low index crystallographic pole and was found to eliminate the region at the center of the pole where few ions were detected. A more refined version of this model indicated that the shape of the emitter evolved during the removal of atoms from the surface and that the equilibrium shape of the specimen surface is not a hemispherical cap. In addition, the simulated image on the single atom detector exhibits dark lines along the major zone lines, as shown in Fig. 5.14 [15]. This study also indicated that the ion trajectories should be calculated on a near atomic scale.

Field evaporation images and these types of simulations have also demonstrated that there are significant differences between the trajectories of the surface atoms and the image gas ions. In the case of the surface ions, the ion trajectories are strongly influenced by the neighboring atoms and the resulting local field distribution. As image gas atoms are ionized further from the surface of the specimen, they are less influenced by the details of the surface structure. Field evaporated ions also experience slightly different trajectories than image gas ions used to produce the field ion image because the ions are field evaporated with a pulsed field. The magnitude of this effect

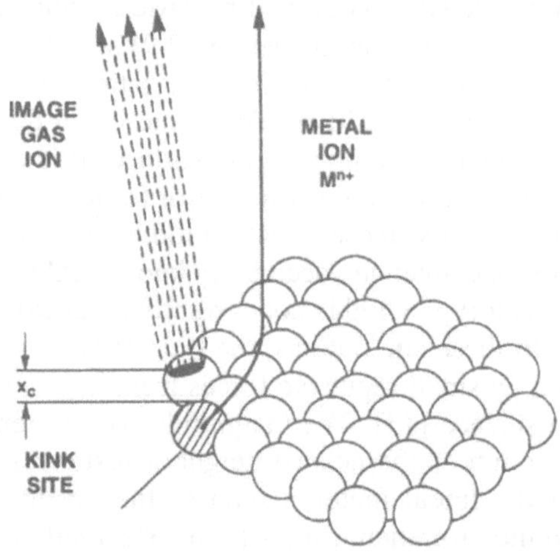

Fig. 5.13. Schematic diagram of the difference in the ion trajectory of the atom leaving the specimen surface and those of the image gas atoms used to form the field ion image. As the ion leaves the surface, it may move laterally before being projected radially. Courtesy A. R. Waugh, University of Cambridge [11].

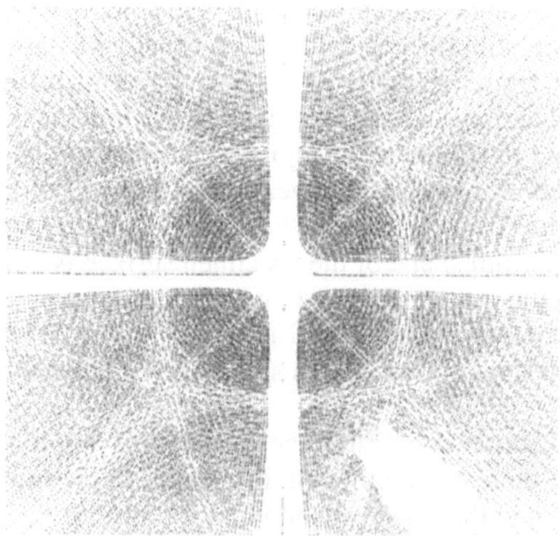

Fig. 5.14. Simulated field evaporation image produced by modelling the trajectories of the ions showing the presence of trajectory aberrations at zone lines and at the centers of poles. Courtesy F. Vurpillot, A. Bostel and D. Blavette, Universty of Rouen [14].

depends on the mass of the ions, the applied voltage and the pulse fraction since it is a function of the time taken for the ions to traverse the distance to the field free region. If the field strength drops suddenly before the ion reaches the field free region because of the termination of the high voltage evaporation pulse, the transverse component of the field decreases. Therefore, the deflection of the ion from its initial radial trajectory is less that expected with a constant field.

5.5.4 Reflectron lens

The incorporation of a reflectron lens into the mass spectrometer produces a small systematic position error due to chromatic aberration. The aberration arises due to the variations in the ion energies produced by incomplete acceleration during the field evaporation pulse. Since ions with lower energy penetrate a shorter distance into the lens, they spend less time in the lens than ions with higher energy. Therefore, ions with lower energies strike the detector earlier than they would without energy compensation. Since the lateral component of the velocity is not affected, the ion travels a shorter distance. The difference in the distance traveled by an ion with energy E and velocity v and one with an energy of E-ΔE and velocity v-Δv is given by [16]

$$\frac{\Delta x}{x} = \frac{\Delta v_x}{v_x} \approx \frac{1}{2}\frac{\Delta E}{E}, \qquad 5.39$$

where x is the lateral distance traveled by the ion of energy E. The result of this chromatic aberration is that the ions of different masses are shifted laterally by small amounts (equivalent ~0.1 to ~0.2 nm on the specimen surface) depending on the flight times of the ions, the applied voltage and the pulse fraction. The corrected position, x', is given for $\beta_p < 1$ by

$$x' = x\left(1 - \frac{1}{2}\frac{\Delta E}{E}\right) = x\left(1 + \frac{1}{2}\frac{(1-\beta_p)V_{pulse}}{(V_{dc}+V_{pulse})}\right), \qquad 5.40$$

where the pulse factor parameter β_p has been empirically found to be given by [16]

$$\beta_p = 0.96 - 0.164\,\log_{10}\left(\frac{t_c}{t_p}\right)^2, \qquad 5.41$$

where t_c is the time taken to traverse the distance, d_c, from the specimen to the counter electrode ($t_c = t\, d_c/d$), d is the specimen-to-detector distance, t is the flight time of the ion from the specimen to the detector, and t_p is the duration of the high voltage pulse.

5.5.5 Local magnification and trajectory aberrations

The most significant factor that influences the spatial resolution is the difference in the local magnification of different phases [17]. This local magnification effect arises from trajectory variations, as discussed in §3.2.9. The presence and the magnitude of the differences in local magnification may be evaluated by counting the total number of atoms in similar size volumes in different regions of the analyzed volume. Atom probe experiments on two-phase materials have demonstrated that the number of atoms collected per unit area on the single atom detector can vary by factors of between 2 and 5 for different phases. In extremely severe cases, the trajectories of the ions from the adjacent phases may overlap. Since the basic reconstruction algorithm does not currently take these trajectory differences into account, the positions of the atoms in the phases are not correctly located. Inaccuracies in the spatial extent of the phases may result. However, an approximate correction may be made to the size of an object, such as a precipitate, from the measured difference in the numbers of atoms per unit volume between the matrix and the object. The corrected size may be approximately evaluated from

$$r_p' = r_p \sqrt{\frac{n_o}{n_m}}, \qquad 5.42$$

where r_p is the size of the object, and n_o and n_m are the number of atoms per unit volume in the object and matrix, respectively. This method assumes that the compression or expansion of the trajectories does not influence the scale in the z direction. In practice, this assumption is not valid since the reconstruction algorithm automatically increments the z direction for each ion collected based on an average atomic volume. Therefore, when a phase is encountered in which more atoms are detected per unit area than the surrounding matrix, the z direction is increased to maintain a constant atomic volume.

Trajectory aberrations have also been observed in field evaporation images of grain and other types of boundaries, §3.2.9. Therefore, the spatial accuracy of the atom positions in these regions is lower than in the interior of the grain. Estimates of the extents of the trajectory aberrations at grain boundaries are typically of the order of 0.5 to 1 nm.

Asymmetrical specimens, which can result from complications during specimen preparation such as preferential etching, may exhibit non-uniform magnification, which affects the accuracy of the reconstruction. This non-uniformity may be observed in the field ion image by an asymmetrical image with elongated image spots.

5.5.6 Other factors affecting reconstruction

The positions of the atoms as they leave the specimen during analysis may not correspond precisely to their atomic positions in the bulk material. The specimen is subject to stresses arising from the cryogenic temperatures, producing a contraction of the atomic lattice, and the field applied to the specimen, producing an expansion. Stress relaxation at the surface may also effect a change in atomic positions, especially for atoms significantly different in size. However, these effects are small and are normally neglected.

5.6 Detection Efficiency

Not all atoms field evaporated from the specimen are detected at the single atom detector. Some ions are lost because they strike or are deflected by the mesh on the entrance and exit electrode to the reflectron lens. This reduces the detection efficiency by the square of the transparency of the mesh as the ions pass through the mesh twice. Additionally, as discussed in §4.3.2, not all atoms that strike the single atom detector are detected. The detection efficiency of the single atom detector is typically 55 to 65% without a biased mesh and up to ~95% with a biased mesh. In addition, pile-up due to multiple ions impinging the detector at the same time has been discussed in §5.4.3.

In addition to these factors, the positions and identities of some ions that are detected by the single atom detector cannot be properly determined for a variety of reasons. The magnitude of this loss varies with the different types of three-dimensional atom probes. In the position-sensitive atom probe and optical atom probe, ions that strike the detector at different positions on the same field evaporation pulse cannot be identified. This multiple hit also occurs but to a much lower extent on the individual anodes in the optical position-sensitive atom probe and the optical tomographic atom probe. In the tomographic atom probe, ions with electron clouds that overlap significantly can be difficult to deconvolute. Lowering of the average field evaporation rate can reduce the magnitude of this problem to some extent. However, if the evaporation field is lowered too much, hydrogen ions can be formed. Preferential evaporation may also be increased due to the presence of hydrogen. The formation of hydrogen ions generally increases the number of pulses on which a hydrogen ion and a metal ion are field evaporated and therefore cancels the beneficial effect. In the tomographic atom probe, ions with electron clouds that extend only over one or two anodes cannot be accurately positioned. In all these cases, the mass-to-charge ratios of the un-positioned ions are determined and can be used in some types of analyses.

The overall detection efficiency is the product of these terms. Typically, overall detection efficiency is given by

$$\zeta = \zeta_d \tau_r^2 \zeta_p \cdots , \qquad 5.43$$

where ζ_d is the detection efficiency of the single atom detector, τ_r is the transmission efficiency of the reflectron, and ζ_p is the positioning efficiency of the ions detected on the single atom detector.

One of the important consequences of this loss is that nearest neighbor ions in the data are not necessarily nearest neighbor atoms in the specimen.

5.7. Specimen Rupture or Failure

An atom probe experiment is normally terminated when the voltage or pulse fraction limit of the instrument is reached, the specimen fractures (flashes) in the microscope, or when a sufficient number of ions have been collected. Specimens may also undergo smaller non-catastrophic ruptures during the course of the experiment. These minor ruptures are manifested in sudden increases in the applied voltages if the control computer is automatically maintaining a constant evaporation rate. Examination of the increase in the applied voltages as a function of the number of ions collected provides an effective method to detect the occurrence of these minor ruptures, as shown in Fig. 5.15a. Data sets that exhibit these sudden increases in voltages should generally be split into separate sections, as there is likely to be a spatial discontinuity in the data. A rough estimate of the amount of material removed in the rupture, in terms of the number of atoms lost, may be made by sliding the portion of the data obtained after the rupture along the horizontal axis until it matches the continuation of the original curve. In contrast, a typical specimen with small taper angle exhibits a gradual increase in the number of ions collected as a function of applied voltage, as shown in Fig. 5.15b. These plots of voltage versus number of ions collected also exhibit changes in the gradient when phases with different evaporation fields are encountered during the analysis, as shown in Fig. 5.15c. Ruptures may generate sufficient heat to melt, dissolve precipitates or otherwise affect the microstructure [18]. Therefore, the data collected immediately after a rupture may not be representative of the microstructure.

References

1. D. Blavette, A. Bostel, J. M. Sarrau and J. Gallot, *Rev. Phys. Appl.*, **18** (1982) 435.
2. D. Blavette, Doctoral thesis, 1981, University of Rouen, France.
3. M. K. Miller, A. Cerezo, M. G. Hetherington and G. D. W. Smith, Atom Probe Field Ion Microscopy, Oxford University Press, Oxford, UK, 1996, p. 86 and associated references therein.

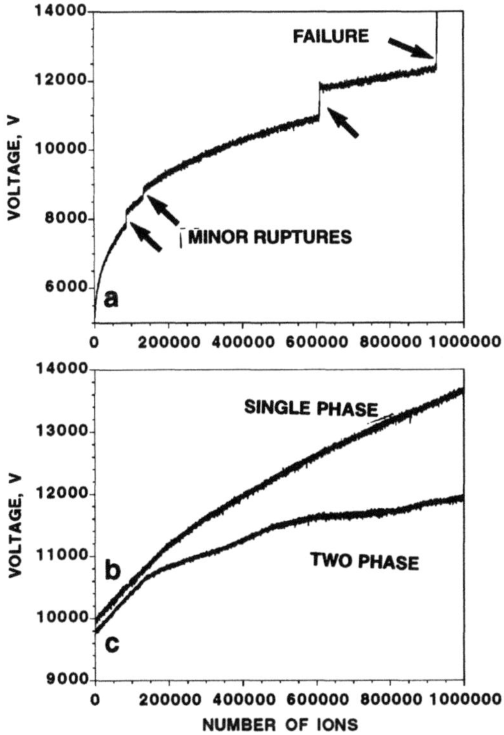

Fig. 5.15. The number of ions collected at the single atom detector as a function of the applied voltage to the specimen. a) Abrupt discontinuities in this curve provides a means to detect minor specimen ruptures. b) Single phase material showing a gradual increase in voltage. c) Changes in gradient are generally an indication of encountering a second phase during the analysis.

4. T. T. Tsong, Y. S. Ng and S. V. Krisnaswarmy, *Appl. Phys. Lett.*, **32** (1978) 778.
5. A. Cerezo, G. D. W. Smith and A. Waugh, *J. de Phys.*, **45-C9** (1984) 329.
6. A. Menand, T. Al-Kassab, S. Chambreland and J. M. Sarrau, *J. de Phys.*, **49-C6** (1988) 353.
7. S. J. Sijbrandij, private communication.
8. B. Deconihout, A. Bostel, P. Bas, S. Chambreland, L. Letellier, F. Danoix and D. Blavette, *App. Surf. Sci.*, **76/77** (1994) 145.
9. P. Bas, A. Bostel, B. Deconihout and D. Blavettte, *Appl. Surf. Sci.*, **87/88** (1995) 298.
10. T. T. Tsong, *Surf. Sci.*, **70** (1978) 211.
11. S. S. Babu, private communication.
12. A. R. Waugh, E. D. Boyes and M. J. Southon, *Surf. Sci.*, **61** (1976) 109.

13. A. J. W. Moore, *Philos. Mag. A*, **43** (1981) 803.
14. F. Vurpillot, A. Bostel, A. Menand and D. Blavette, *Eur. Phys. J. Appl. Phys.*, **6** (1999) 217.
15. F. Vurpillot, A. Bostel and D. Blavette, *J. Microsc.* **196** (1999) 332.
16. A. Cerezo, T. J. Godfrey, S. J. Sijbrandij, G. D. W. Smith and P. J. Warren, *Rev. Sci. Instrum.*, **69** (1998) 1.
17. M. K. Miller and M. G. Hetherington, *Surf. Sci.*, **246** (1991) 442.
18. S. S. Brenner, J. Kowalik and H. Ming-Jian, *Surf. Sci.*, **246** (1991) 210.

Chapter 6

Data Representations and Analysis

Many different methods have been developed to represent and analyze the data obtained in the three-dimensional atom probe [1-4]. The mass spectrum and the plots of the number of ions collected as a function of the applied voltage have been discussed in previous chapters. Other diagnostic displays are available during the experiment to ensure that the instrument is operating correctly, as discussed in Chapter 4. In this chapter, the standard methods of representing and analyzing the three-dimensional data are presented. These methods are usually performed in a high performance workstation after the experiment has been completed. Some of the simpler representations may be used on line to monitor the progress of an analysis with respect to the microstructural features present.

6.1 Visualization and Analysis Methods for Individual Atoms

In this section, the methods available to visualize individual atoms and to measure parameters from the coordinates of individual atoms are discussed.

6.1.1 Atom or dot map representations

The simplest and most intuitive representation of the three-dimensional data is the atom, or dot, map. In this type of representation, a dot or a small sphere is plotted at the x, y, and z coordinates of the atom. The size of the sphere is usually selected to be approximately equivalent to the magnified diameter of the atom. These volumes are typically manipulated in the computer in real time so that the local arrangement of the atoms may be visualized. Single pixel dots are generally used to improve the speed of the display, particularly during extensive manipulation of the volume, due to the additional length of time required to render a sphere. Although it is possible to display all the atoms in the volume simultaneously, the representation is often more informative when only one or a small subset of atom species are displayed at a time. Examples of atom maps are shown in Fig. 1.16. Additional examples are shown elsewhere in this monograph. Different atom types may be visualized simultaneously with the use of color-coded dots or small spheres and/or with the use of different diameter spheres, as shown in Plates I –V and VII-X.

This type of representation is extremely effective when the matrix concentration of the solute of interest is relatively low and there is a high concentration of solute in the feature of interest. Atom map representations are useful to visualize features such as clusters, precipitates or solute segregation to grain boundaries, interphase interfaces or dislocations. In ordered alloys, atom maps may be used to detect visually the site preference of a solute.

6.1.2 Defining features of interest

In cases where there is a relatively small but significant solute concentration in the matrix, the visibility of features may be improved by rendering only those atoms that are within a certain distance of another solute atom of the same type or group of types. The magnitude of the typical distance is somewhat dependent on the solute concentrations in the features and the matrix and are generally of the order of a few nearest neighbor distances (i.e., ~0.3 to ~0.7 nm). Since the solute atoms in the matrix are further apart than this distance, the solute in the matrix is eliminated from the visualization. An example of this method applied to an iron-copper-nickel alloy is shown in Fig. 6.1. Atom probe analysis of this material revealed that 0.54±0.08% Cu remained in solid solution after the aging treatment of 100 h at 400°C. In these atom map representations, all the copper atoms in the volume of analysis are plotted in Fig. 6.1a and only those copper atoms that were within 0.5 nm of another copper atom are displayed in Fig. 6.1b. The majority of the copper atoms in solid solution in the matrix were eliminated in Fig. 6.1b.

These atoms may also be grouped into specific features from the distribution of the atom locations and the number of atoms associated with each feature may be estimated. These data may then be used to determine the center of mass (§6.1.3) and the radius of gyration (§6.5.2) of each feature.

The atoms in the feature may be used to define its extent by creating a three-dimensional envelope. The envelope is created with the use of a fine grid. A two-dimensional section through a typical grid is shown in Fig. 6.2. For each atom in the feature, the cell in the grid in which the atom is located is marked. Cells on each row or column of the grid that are surrounded on at least three sides in two dimensions (or five sides in three dimensions) by

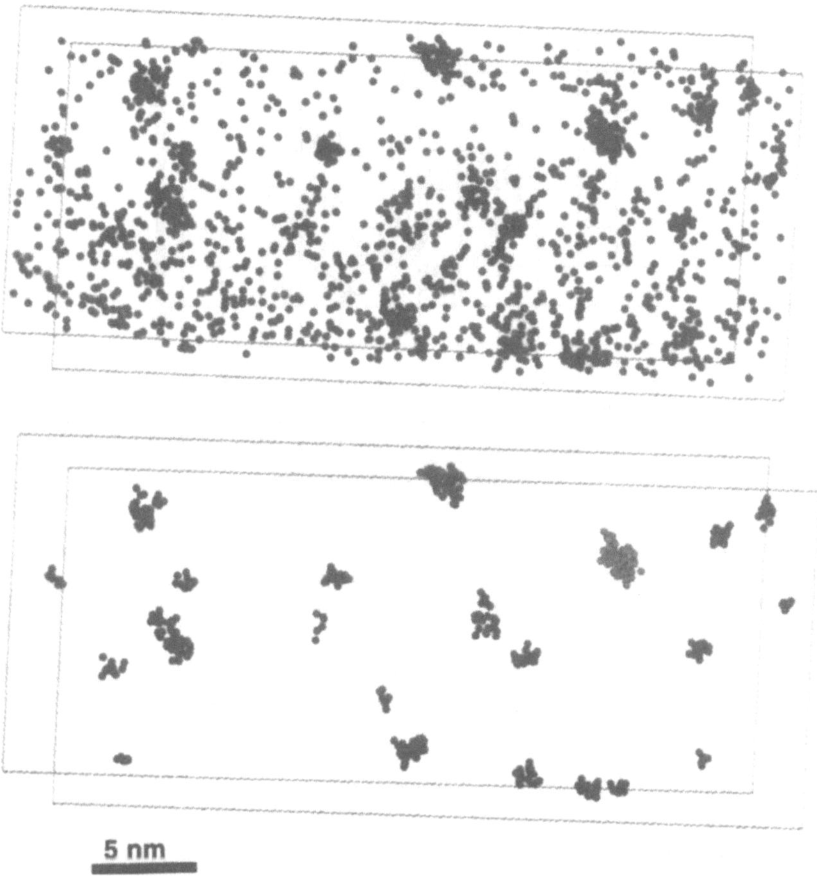

Fig. 6.1. a) Atom map of the copper atoms in an Fe-1.1 at.% Cu – 1.4% Ni alloy that was aged for 100h at 400°C. b) Atom map in which only the copper atoms that are within a specified distance (0.5 nm) of another copper atom are displayed. This method eliminates the copper (c=0.54±0.08 at. % Cu) that is in solid solution in the matrix and therefore enhances the prominence of the copper-enriched precipitates.

marked cells are then filled in. The outermost cells then define the envelope. The extent of the feature may be estimated from the distances between the cells on the periphery of the feature. The composition of the feature can also be calculated from the numbers of all the different types of atoms within that envelope.

A simple estimate of the minimum size of the feature may be made from the number of atoms associated with the feature, n, the atomic volume of the

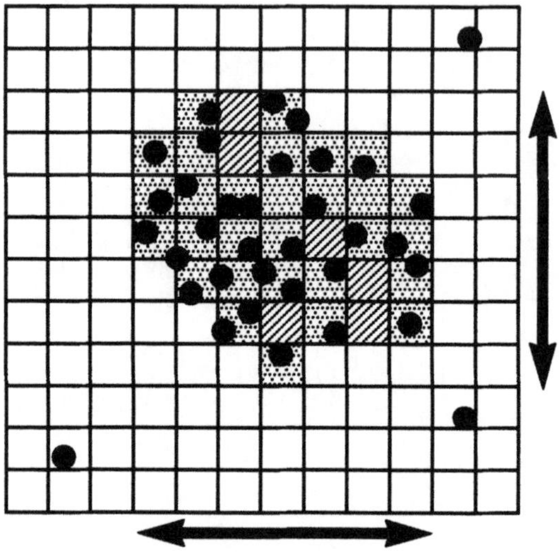

Fig. 6.2. Creation of a three-dimensional envelope from the coordinates of the atoms. In this two-dimensional section of the three-dimensional data, the cells in which atoms (black dots) are detected have been marked by the grey shading and those that are filled in by interpolation have the hatching pattern.

atoms, Ω, and the detection efficiency of the single atom detector, ζ. For example for a spherical particle, the radius, r_p, may be expressed as

$$r_p = \sqrt[3]{\frac{3\,n\,\Omega}{4\pi\,\zeta}}. \qquad 6.1$$

6.1.3 Center of mass

If the coordinates of all the atoms in a feature are known, the center of that feature may be determined. This parameter is required for a number of different types of analyses, as discussed in the following sections.

The center of mass is the point in a body or system of bodies at which the entire mass may be considered as being concentrated. In the case of an atom probe analysis, the center of mass $(\bar{x}, \bar{y}, \bar{z})$ of a feature, such as a precipitate, is given by

$$\bar{x} = \frac{\sum_{i=1}^{n} x_i\, m_i}{\sum_{i=1}^{n} m_i}, \quad \bar{y} = \frac{\sum_{i=1}^{n} y_i\, m_i}{\sum_{i=1}^{n} m_i}, \quad \text{and} \quad \bar{z} = \frac{\sum_{i=1}^{n} z_i\, m_i}{\sum_{i=1}^{n} m_i}, \qquad 6.2$$

where x_i, y_i and z_i are the spatial coordinates of each atom, m_i are their masses, and n is the number of atoms in the feature. If all atoms are the same species (i.e., the same mass), this reduces to [5]

$$\bar{x} = \frac{\sum_{i=1}^{n} x_i}{n}, \quad \bar{y} = \frac{\sum_{i=1}^{n} y_i}{n}, \quad \text{and} \quad \bar{z} = \frac{\sum_{i=1}^{n} z_i}{n} . \qquad 6.3$$

6.1.4 Local compositions

The simplest method of local composition determination is to estimate the composition of a small selected volume at a specific location in the analyzed volume. The location of this volume is selected either manually by the user or from other parameters such as the center of mass, as discussed in §6.1.3. The size of the volume may be selected to match the size of the object under analysis. In some cases, such as for a matrix analysis, it may encompass the entire volume of analysis. The concentration in atomic fraction of a solute in the volume is given by

$$c_i = \frac{n_i}{n_t}, \qquad 6.4$$

where n_i is the number of ions of solute i, and n_t is the total number of ions collected. The estimate of the standard error of this measurement, s, assuming a random solid solution, is given from the binomial distribution by

$$s = \sqrt{\frac{c_i(1-c_i)}{n_t}} . \qquad 6.5$$

The standard error approximates to

$$s = \frac{\sqrt{n_i}}{n_t}, \qquad 6.6$$

for large samples and low solute concentrations.

The variations in the local concentrations in the volume of analysis may also be examined by stepping a small volume through the atom-by-atom data and determining the composition of each volume from the relative numbers of atoms of each type that are encompassed in that selected volume. This selected volume is generally either a small cube or a sphere. The step size can be the same size or smaller than the extent of the selected volume. The size of this selected volume is a compromise between maximizing the number of atoms from, and not exceeding the extents of, the smallest feature of interest in the analyzed volume. However, the selected volume should also be large enough to contain a sufficient number of atoms for a reliable compositional determination to be made. After the compositions of these volumes have been determined, they may be sorted according to the concentrations of one or more elements in the alloy. The location of the selected volumes that satisfy specific criteria, such as the selected volumes that exceed a threshold concentration, may be superimposed on an atom map. For example, the 2 x 2 x 2 nm selected volumes with the highest chromium concentrations in a spinodally decomposed iron-chromium alloy are shown in Fig. 6.3. It is evident that these volumes correspond to regions of the chromium-enriched α' phase.

6.1.5 Composition profiles and radial concentration profiles

One of the most common forms of data analysis is the construction and statistical analysis of a composition profile. These types of profiles may be used to establish the presence of second phases, to estimate the concentration differences and solute partitioning between phases, and to detect solute segregation to interfaces. The ability to orient the axis of the composition profile with respect to a feature of interest and to vary the lateral area over which the composition data is averaged are major advantages of the three-dimensional atom probe over the classic atom probe.

The first step in the process to create a composition profile is to select the volume within the three-dimensional data from which the composition profile is desired. The shape of this selected volume is usually either a cylinder or a rectangular block depending on the type of feature under analysis and its shape in the analyzed volume. The size and position of this selected volume are then carefully adjusted by the user until the desired volume is enclosed. In practice, this operation is generally performed by dragging the edges of the selected volume that is superimposed on one of the standard representations, such as an atom map or isoconcentration surface. Once the volume has been selected and correctly positioned, the atom-by-atom data within the selected volume are divided into small slices perpendicular to one of the axes of the selected volume. The size of the slices may be defined in terms of either a fixed number of atoms or distance. Larger slices produce less statistical scatter in the measurements but reduce the spatial resolution as indicated by eqn. 6.5. The concentrations of all the elements present in each slice are then determined based on the number of atoms of each elemental species with the use of eqn. 6.4. An example of a composition profile across an interface is shown in Fig. 6.4.

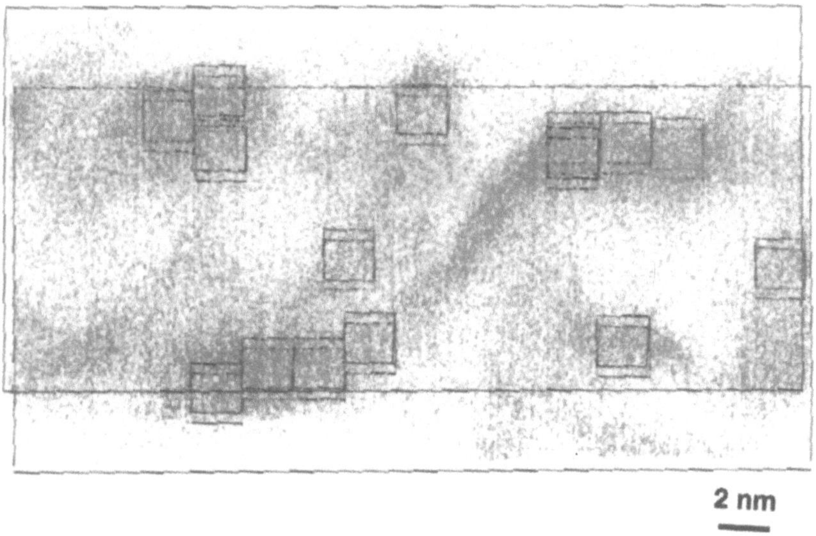

Fig. 6.3. Atom map of an Fe-32% Cr alloy aged for 10,800 h at 470°C with the selected 2 x 2 x 2 nm volumes containing the highest concentrations of chromium superimposed.

Another form of composition profile is the radial concentration profile. In a radial concentration profile, the concentrations of the various types of toms within thin spherical shells are determined as a function of distance from the origin of the feature. The origin of the analysis is usually the center of mass of a spherical object, such as a precipitate. An example of a radial concentration profile from a copper-enriched precipitate is shown in Fig. 6.5. In this method, the error on the measurements decreases with distance from the origin since the spherical shells contain more atoms.

6.1.6 Number density

The number density of a distribution of discrete particles may be calculated from the number of particles observed, N_p in the volume analyzed, v, i.e.,

$$N_v = \frac{N_p}{v}. \tag{6.7}$$

In principle, the volume could be estimated from the products of the extents of the x, y and z directions of the analyzed volume. However, the extents of the x and y dimensions of these data generally increase during the analysis

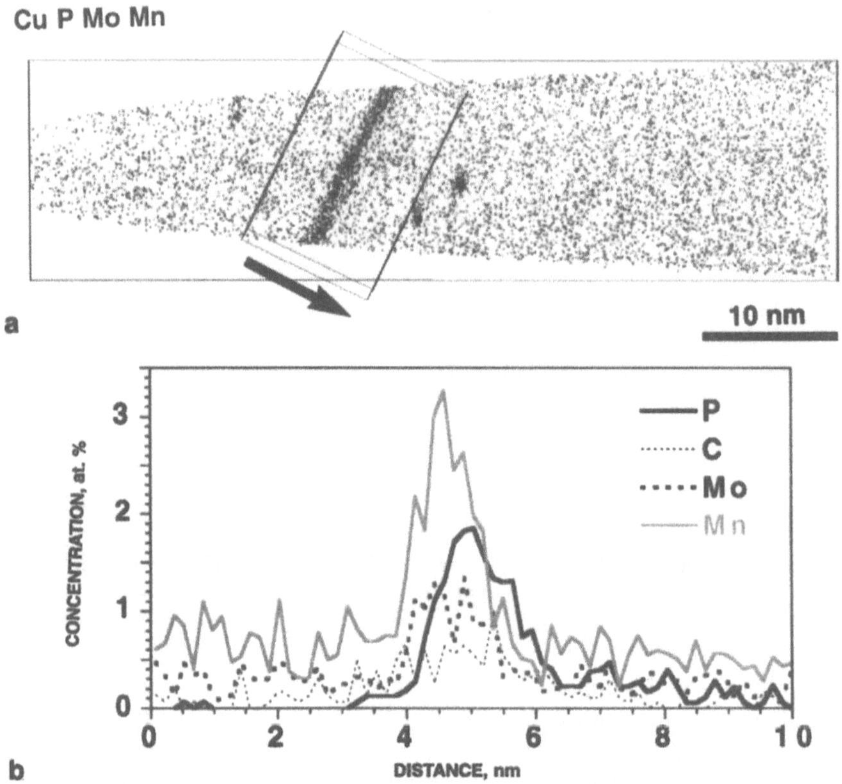

Fig. 6.4. Atom probe analysis in the vicinty of an interface in a pressure vessel steel. a) atom map; b) composition profile across the interface indicated in (a). Enrichments of P, C, Mo and Mn are evident.

due to the blunting of the specimen, making a reliable estimate difficult. Therefore, the largest box that may be inscribed within the data is generally selected to perform this determination.

Alternatively, the volume may be estimated from the total number of atoms detected in the volume, n, the average atomic volume, Ω, and the detection efficiency of the single atom detector, ζ. Therefore, the number density is given by

$$N_v = \frac{N_p \zeta}{n \Omega}. \tag{6.8}$$

If some of the particles are not fully contained in the volume and some clipping has occurred, these simple estimates will overestimate the actual

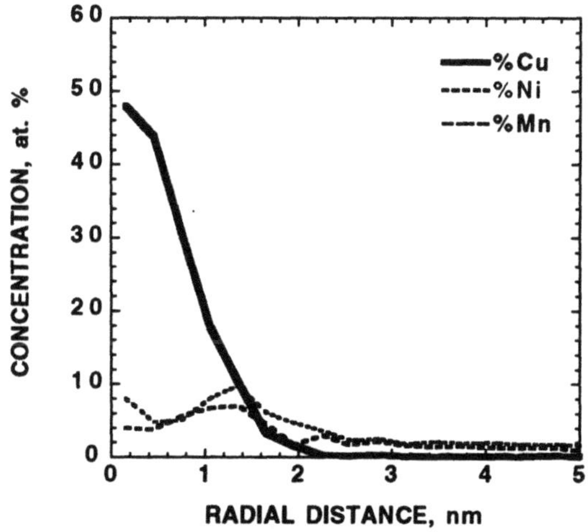

Fig. 6.5. Radial composition profile from the center of a copper-enriched precipitate to the surrounding matrix in a neutron irradiated pressure vessel steel weld showing significant enrichments of nickel and manganese both in the precipitate and at the precipitate-matrix interface.

number density. Two methods are commonly used to overcome this bias. The first method is to include only those precipitates whose centers lie in the volume. Unfortunately, it is not always possible to estimate the precise center of a clipped particle. The second method is to randomly select one corner of the volume and include only those precipitates that are not clipped by the three sides that intersect at that corner, as shown in Fig. 6.6. This process can be repeated at each of the eight corners and an average value calculated.

6.1.7 Fast Fourier transform

In three-dimensional atom probe analyses that are performed close to crystallographic poles, the planes of atoms may be evident in the atom map reconstructions. The Fast Fourier transform [6] may be applied to locate and visualize additional sets of planes in the data [7,8]. Each set of crystallographic planes resolved in the atom probe analysis is represented by a spot in the Fourier transformed data, as shown in Fig. 6.7. By performing an inverse Fourier transform with intensity from one of these spots, the corresponding series of planes can be visualized in the three-dimensional data. This information may be used to estimate the accuracy of the reconstruction from the angles and interplanar spacing of the spots and, if necessary, to adjust the parameters used. In principle, if the analysis contains a grain boundary or other type of interface, the orientation relationship

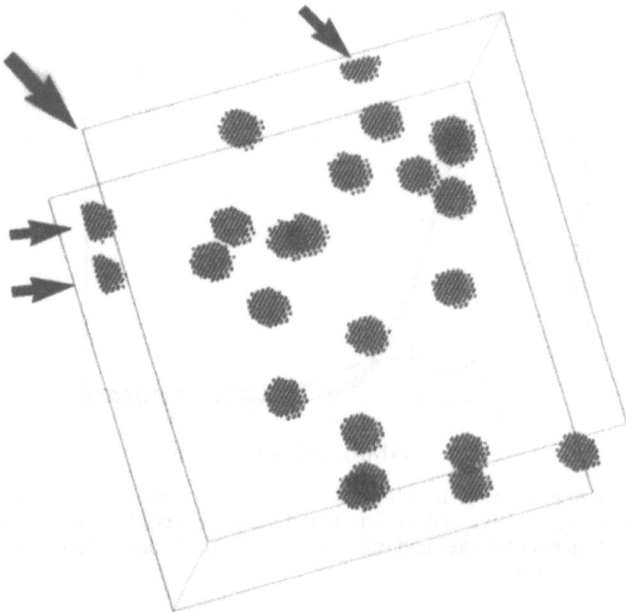

Fig. 6.6. Atom map of simulated 2-nm-diameter copper precipitates. The copper precipitates that are cut by the sides of the volume of analysis that intersect at the marked corner are not included in number density calculations to prevent bias.

between the grains may be estimated. It has also been suggested that this method may provide a means of placing the atoms on their exact lattice sites, assuming that the crystal structure is known [9].

6.2 Smoothing data

Several of the methods of analysis require that the ion-by-ion data be converted into a three-dimensional array of concentrations. The methods to perform this conversion are described in this section.

6.2.1 Method to place the ion-by-ion data on a grid

In order to convert the ion-by-ion data into composition maps, a three-dimensional grid of data locations is defined to accumulate the number of atoms of each element or range [2]. This process is sometimes referred to as "binning the data". Each data location in this grid corresponds to a small equiaxed volume, or volume element (or voxel), in the specimen. The spatial extent of these volume elements in the specimen depends on the type of data

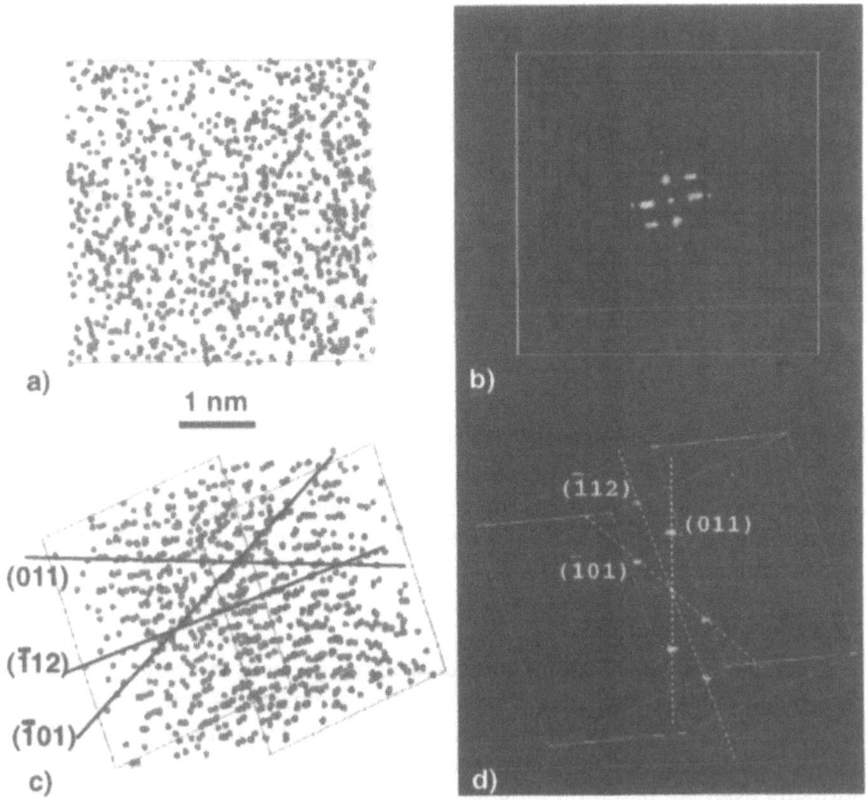

Fig. 6.7. a) Atom map of a 4 nm cube of a tungsten specimen. b) FFT showing the presence of spots in the reciprocal lattice corresponding to the sets of planes. c) Rotated atom map revealing three sets of atom planes. d) Reciprocal lattice rotated so that the 6 major spots lie in the plane of the image. Courtesy P. J. Warren, University of Oxford [7].

analysis to be performed. Smaller grid spacings retain the spatial resolution whereas larger grid spacings have more atoms assigned to each data location and therefore exhibit smaller fluctuations in the composition determinations. In this method, the number of atoms assigned to each data location in the grid varies and some locations may not contain any atoms (i.e., a sparse matrix), especially when small grid spacings are used. The identity of each ion is determined, as discussed in §5.4, and is then assigned to a specific data location in the grid based on its x, y and z coordinates, as shown schematically in Fig. 6.8. After all ions have been assigned, the concentrations of all the elements within each data location in the grid may be determined from eqn. 6.4. Although the grid is stored in memory, the

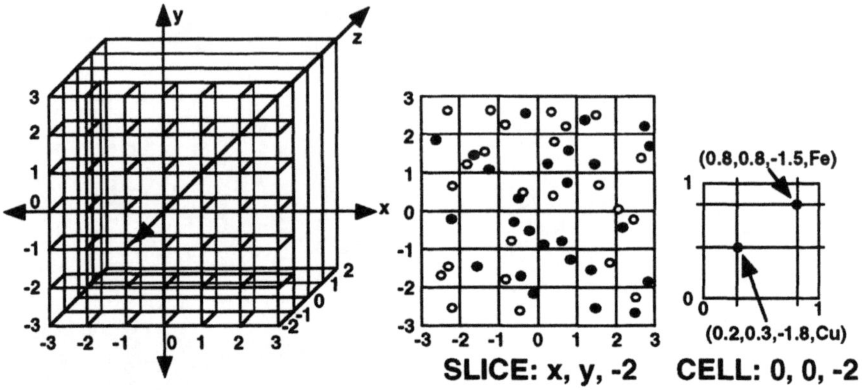

Fig. 6.8. Placing the ion-by-ion data onto a three-dimensional grid of data locations.

operator may interrogate the composition of each data location in the grid by positioning a pointer on a selected voxel in a three-dimensional representation of the grid.

6.2.2 Smoothing methods

Since the number of atoms in each volume element is limited, there are significant fluctuations in the measured concentrations due to statistical sampling. Therefore, it is often desirable to smooth the statistical fluctuations before proceeding with the data analysis [10,11]. The smoothing methods normally used are based on moving averaging techniques. Therefore, the 2-, 3- and 4-point smoothing methods use kernels containing 2^3, 3^3, 4^3 values, respectively to determine the average concentration. Higher order kernels are typically not used, as they are computationally intensive and smooth the data excessively. For a 2-point smoothing, the output value is given by the sum of the products of the original data, $c_{h,i,j}$ and the corresponding multiplying factor in the kernel, $k_{u,v,w}$, i.e., $o_{h,i,j} = c_{h,i,j}k_{u,v,w} + c_{h+1,i,j}k_{u+1,v,w} + c_{h,i+1,j}k_{u,v+1,w} + c_{h+1,i+1,j}k_{u+1,v+1,w} + c_{h,i,j+1}k_{u,v,w+1} + c_{h+1,i,j+1}k_{u+1,v,w+1} + c_{h,i+1,j+1}k_{u,v+1,w+1} + c_{h+1,i+1,j+1}k_{u+1,v+1,w+1}$, where h, i, and k are the data locations in the grid and u, v and w are the positions of the multiplying factors in the kernel. This pattern is extended with additional terms for 3- and 4-point smoothing schemes. An unweighted two-dimensional variant of this scheme is demonstrated in Fig. 6.9. The multiplying factors for the unweighted and two differently weighted 3-point kernels are shown in Table 6.1. In the unweighted case, all the multiplying factors are identical and correspond to the reciprocal of the number of values in the kernel (i.e., 2-point $k_{u,v,w}$ = 0.125, 3-point $k_{u,v,w}$ =

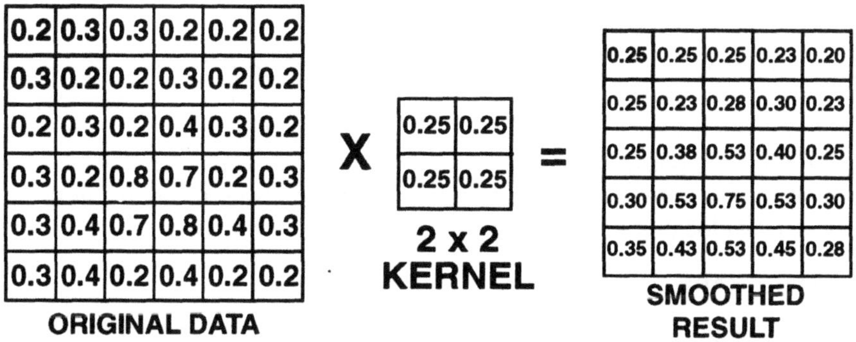

Fig. 6.9. Smoothing the three-dimensional composition data. The concentrations are expressed in atomic fraction. The values in the smoothed data are the sum of the products of the original data by the corresponding values in the kernel, e.g., for the 4 data in the shaded region in this two-dimensional simplification, the smoothed value is 0.2 x 0.25 + 0.3 x 0.25 + 0.3 x 0.25 + 0.2 x 0.25 = 0.25.

Table 6.1 Multiplying factors for three different types of 3-point smoothing kernels.

	-1,y,-1	0,y,-1	1,y,-1	-1,y,0	0,y,0	1,y,0	-1,y,1	0,y,1	1,y,1
Unweighted ($1/3^3 = 0.037$)									
x,-1,z	0.037	0.037	0.037	0.037	0.037	0.037	0.037	0.037	0.037
x,0,z	0.037	0.037	0.037	0.037	0.037	0.037	0.037	0.037	0.037
x,1,z	0.037	0.037	0.037	0.037	0.037	0.037	0.037	0.037	0.037
Weighted - cone									
x,-1,z	0.016	0.034	0.016	0.034	0.058	0.034	0.016	0.034	0.016
x,0,z	0.034	0.058	0.034	0.058	0.116	0.058	0.034	0.058	0.034
x,1,z	0.016	0.034	0.016	0.034	0.058	0.034	0.016	0.034	0.016
Weighted - sine									
x,-1,z	0.017	0.036	0.017	0.036	0.058	0.036	0.017	0.036	0.017
x,0,z	0.036	0.058	0.036	0.058	0.082	0.058	0.036	0.058	0.036
x,1,z	0.017	0.036	0.017	0.036	0.058	0.036	0.017	0.036	0.017

0.037 and 4-point $k_{u,v,w} = 0.0156$. In the weighted cases, the magnitude of the multiplying factor is higher than the unweighted case in the center of the kernel and the multiplying factors decrease with distance from the central value. This type of kernel tends not to degrade the sharpness of interfaces between phases to the same extent as unweighted kernels. Symmetrical kernels, in which the magnitude of the multiplying factor is the same at all similar distances from the center of the kernel, are normally used. However, asymmetrical kernels, such as edge filters, may be used to emphasize special types of interfaces, such as planar interfaces. The resulting average concentration is positioned at the center of the kernel, which corresponds to a data location in cases where an odd number of points is used and to a position between data locations where an even number of points is used.

6.3 Data Representations

In this section, some of the common techniques that are used to visualize variations in local composition from the three-dimensional binned data (§6.2) are described. These representation methods may be used individually or in combination. Multiple representations of the same data set either in one combined representation or separately in different windows can often provide a clearer impression of the microstructure. In addition to the methods presented in this section, the composition profile and the radial concentration profile are described in §6.1.5.

6.3.1 Opacity representations

One of the original techniques to visualize three-dimensional data was the opacity method. The first step in this method is to determine, and optionally smooth, the compositions of all the volume elements in the analyzed volume, as discussed in the previous section. Each data location in the grid is then assigned an opacity factor based on the concentration of one or more of the solutes. Generally, the low concentration regions are selected to be transparent and the highest concentration regions opaque. The opacity factors of the entire volume are then rendered in three dimensions, as shown in Fig. 6.10. The high solute regions in the interior of the analyzed volume may be observed due to the transparency of the low solute regions. This method is rarely used, as other methods of representation are more informative.

6.3.2 X-ray tracer representation

The x-ray tracer is a method of representation that enables the maximum (or minimum) concentrations of features to be visualized and feature to feature comparisons to be made. The first step in the computation is to determine, and optionally smooth, the compositions of all the volume elements in the analyzed volume (§6.2). The maximum (or minimum) solute concentrations of a selected element are determined for each column of data locations in the grid that are perpendicular to one of the x-y, x-z or y-z surfaces of the volume. For example if the y-z plane is selected, the selection

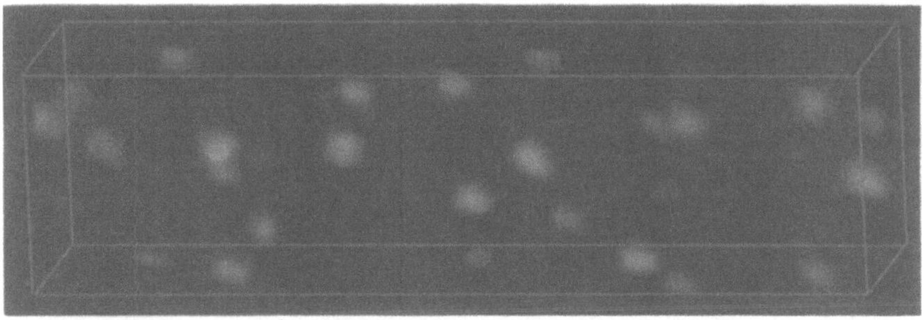

Fig. 6.10. In this opacity representation of copper-enriched precipitates in an Fe-Cu model alloy, the higher concentrations are denoted by more opaque voxels. The highest (lightest) concentration level corresponds to a copper content of 65%. From a collaboration with P. Pareige and D. Blavette, University of Rouen, France.

process is performed on columns of data locations along the x direction. The results are then displayed as a two-dimensional color or gray scale projection, as shown in Plate VI. This method is usually effective in detecting and visualizing the presence of, and the variations, in solute concentrations of nanometer scale precipitates.

6.3.3 Arbitrary slicer

The arbitrary slicer is an alternative method to display local variations in concentration in the analyzed volume. Different colors or a gray scale depicts the levels of solute concentrations. After binning and optionally smoothing the data (§6.2), a plane intersecting the concentration data is selected, as shown in Fig. 6.11. The plane may be orthogonal to, or inclined at an arbitrary angle to, the axes of the volume of analysis. The concentrations of the data locations that intersect that plane are displayed. If the plane is orthogonal to one of the major axes of the volume, it is possible to automatically sweep the plane through the entire volume to reveal the variations in concentration. Alternatively, the position of the plane may be manipulated manually.

6.3.4 Isoconcentration surfaces

One of the most effective methods to visualize the size and morphology of features is the isoconcentration surface, often incorrectly referred to as an isosurface. The isoconcentration surface may be regarded as the three-dimensional equivalent to a contour map. In this method, the solute concentration level of one or more elements is selected and the surface that connects that concentration level throughout the data is constructed, as shown in Fig. 6.12. The example shown in Fig. 6.12a is from an Fe-30 at. % Cr alloy

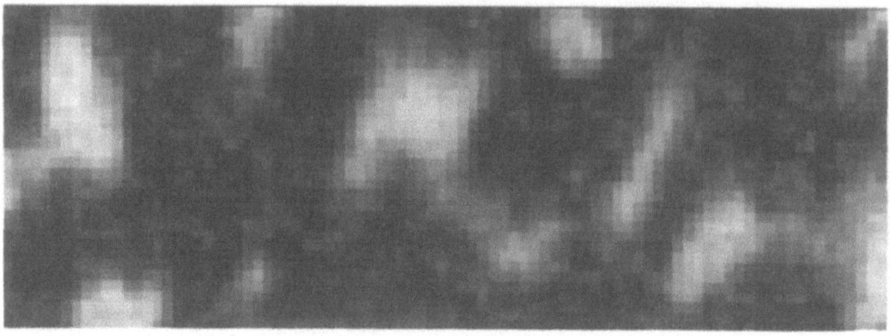

Fig. 6.11. An orthogonal slice through a volume of analysis from an Fe- 32% Cr alloy aged for 10,800 h at 470°C. The gray scale ranges from black (0% Cr) to white (100% Cr) and indicates the local chromium concentration.

aged for 10,800h at 470°C. During this treatment, the alloy undergoes spinodal decomposition and forms a chromium-enriched α' phase and iron-rich α phase. The complex interconnected network structure of the α' phase is evident from the 30% Cr isoconcentration surface that is shown. It is evident that the isoconcentration surface has a brighter exterior surface and a dimmer interior surface so that the direction of the concentration gradient may be represented. In the second example from nickel based superalloy Alloy 718, Fig. 6.12b, two isoconcentration surfaces are shown. The first is based on the levels of iron plus chromium, which partition preferentially to the γ matrix, and depicts the interface between a primary γ'' precipitate and the γ matrix. The second is based on the aluminum level and reveals the presence of small γ' (Ni_3Al) precipitates nucleated on the surface of the primary γ'' precipitate. In some cases, such as complex multi-element precipitates, it may be beneficial to use the concentrations of the elements partitioned to the matrix to form the isoconcentration surface and then reverse the exterior and interior surfaces to reveal the precipitate.

Since isoconcentration surfaces require a significant amount of time to compute, it is common not to use every point in the concentration data but every second or third point, etc. This technique is known as downsizing. Higher downsizes also generally produce slightly smoother surfaces.

When the concentration gradient at an interface is not sharp, the selection of the concentration level for the isoconcentration surface has a strong influence on the extent of the features present. One systematic but arbitrary method to overcome this influence is to use the midpoint of the concentration difference between the phases on either side of the interface.

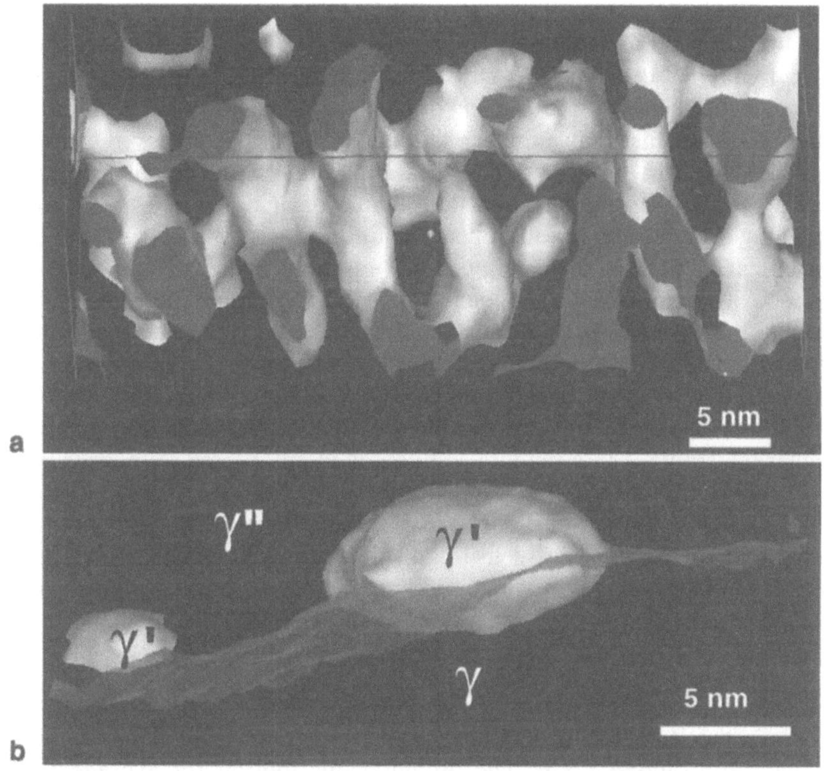

Fig. 6.12. Isoconcentration surface from the analysis of two materials. a) 30% Cr isoconcentration surface from an Fe-30 at. % Cr alloy aged for 10,800h at 470°C. The complex interconnected network structure of the chromium-enriched α' phase is evident from the isoconcentration surface. b) Isoconcentration surfaces from a nickel-based superalloy Alloy 718. The presence of 4-nm-diameter γ' precipitates on the surface of a primary γ'' precipitate is evident from the 8% alumimum isoconcentration surface and the 30% iron+chromium isoconcentration surface, respectively.

6.4 Composition Determinations

In the previous sections, the methods to visualize the data have been presented. However, in most atom probe experiments, quantification of the compositions of the features present, the degree of cosegregation and the levels of solute segregation are desired. The techniques to perform these types of analyses are presented in this section.

6.4.1 Frequency distributions

Information on the solute distribution in a specimen may be determined with a frequency distribution. A frequency distribution is a plot of the total number, or frequency of occurrence, of blocks of data with a given number of solute atoms against that number. Alternatively, the frequency distribution may be plotted in terms of concentration. The frequency distribution is created by dividing the atoms in the volume of analysis into small blocks containing a fixed number of atoms. The dimensions of the block should be approximately equiaxed to maintain the optimum sampling of coexisting phases. An example of a frequency distribution obtained from an iron-chromium alloy in which phase separation into iron-rich α and chromium-enriched α' regions has occurred is shown in Fig. 6.13. The mode (most numerous) and median concentrations may be directly determined from this distribution. The shape of the frequency distribution provides information on whether the solute is in a random solution, is in an ordered state, or is phase separated into high and low solute regions.

The expected number of blocks with i solute atoms, $F(i)$ is given in terms of the probability $P(i, N_b)$ of detecting i solute atoms in blocks containing N_b atoms (i.e., the binominal distribution) by

$$F(i) = n_F P(i, N_b) = n_F \binom{N_b}{i} p^i q^{Nb-1} = n_F \frac{N_b!}{i!(N_b - i)!} p^i q^{N_b - 1}, \qquad 6.9$$

where n_F is the total number of fixed size blocks, p is the probability of success in the Bernoulli trial, which can be estimated from the mean solute concentration, and $q = 1\text{-}p$. The binomial distribution for average solute content of the iron-chromium alloy is also included in Fig. 6.13. The frequency distribution from the experimental data is significantly wider than the binomial distribution indicating that the chromium atoms are not in a random solid solution and phase separation has occurred.

The hypothesis that there is no interaction between the atoms (i.e., the outcome is a result of a Bernoulli trial) may be tested with the use of a χ^2 statistical test. For comparison with the χ^2 distribution, the test statistic X^2 is defined as

$$X^2 = \sum_{i=0}^{n'-1} \frac{(O'(i) - F'(i))^2}{F'(i)}, \qquad 6.10$$

where $F'(i)$ is the expected number of observations of a class, $O'(i)$ is the experimentally observed number of occurrences of a class, and n' is the number of classes $n' \leq N_b$-1. The test statistic X^2 may be compared to the χ^2 distribution as long as the expected number of observations in each class is sufficiently large. To fulfill this requirement, the experimental (and the expected) observations should be combined into classes so that each class

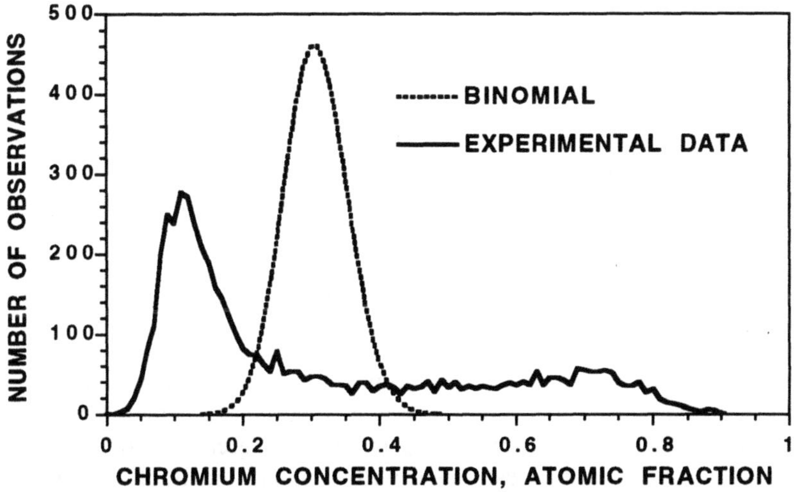

Fig. 6.13. Frequency distribution of the chromium concentrations of 100 ion blocks of atoms from an Fe-32% Cr alloy aged for 10,800 h at 470°C and the binomial distribution for the average chromium content (0.31% Cr). The broadening of the experimental frequency distribution indicates that the alloy has phase separated into iron-rich α regions and chromium-enriched α′ regions.

contains at least 5 observations. For a random solid solution, the null hypothesis is that the observed distribution is consistent with the hypothesized binomial distribution. The null hypothesis is rejected if the value of X^2 exceeds the tabulated α% percentile of χ^2 with $n' - 1$ degrees of freedom, where α is the level of significance chosen for testing the hypothesis.

In materials that have a tendency to order, the width of the frequency distribution may be narrower than that expected from a random solid solution. The reason for this narrowing is that if the solute in a block of atoms deviates from the mean concentration, there will be an increase in the number of atoms on incorrect lattice sites, which would be energetically unfavorable. Therefore, there is a smaller than random chance of detecting blocks of atoms whose compositions deviate from the mean composition.

A broadening of the distribution or the presence of more than one peak indicates the presence of phase separation in the material. A variety of models may be used to describe the microstructure. These models were originally derived to determine the compositions of the phases produced as a result of spinodal decomposition. The experimental frequency distribution may be compared to probability distributions generated by these models in order to estimate the compositions of coexisting phases. These methods are described in the following sections.

6.4.2 P_a method

The solute concentrations in a pair of coexisting phases may be estimated with the P_a or sinusoidal method [12,13]. This method is based on Cahn's sinusoidal or linear theory of spinodal decomposition [14] and assumes that the concentration fluctuations are of the form

$$p_j = c_0 + P_a\left[\frac{2\pi j}{m_d}\right], \quad \{j : 0 \le j \le m_d\}, \tag{6.11}$$

where $2P_a$ is the peak-to-peak amplitude of the spinodal, c_0 is the mean solute concentration and m_d is the discretization of the composition profile. The value of the discretization is chosen to ensure that the estimate is independent of the value of the discretization and m_d is typically greater than 20.

The value of P_a may be estimated with a maximum likelihood method [12,13] or moments estimator [15,16]. For example, the likelihood of obtaining i solute atoms in a block containing N_b atoms is given by

$$P(i, N_b) = \frac{1}{m_d}\sum_{j=0}^{m_d-1}\binom{N_b}{i} p_j^i q_j^{N_b-1}, \tag{6.12}$$

where $q_j = 1 - p_j$. The likelihood is maximized by finding the best fit (i.e., the maximum) of $S = \log|L|$ with respect to P_a, by stepping through all possible values of P_a, $(0 \le P_a \le c_0)$ where

$$S = \log|L| = \sum_{i=0}^{N_b} O(i)\log|P(i, N_b)|, \tag{6.13}$$

and $O(i)$ is the number of times a block containing i solute atoms was observed.

The error in the P_a measurement, σ, is estimated from the variance, σ^2, by

$$\sigma^2(P_a) = \frac{1}{m_d}\frac{1}{\sum_{i=0}^{N_b}\frac{1}{P(i, N_b)}\left(\frac{\partial P(i, N_b)}{\partial P_a}\right)^2}, \tag{6.14}$$

where the partial derivative is given by

$$\frac{\partial P(i, N_b)}{\partial P_a} = \frac{1}{m_d}\binom{N_b}{i} p_j^i q_j^{N_b-1}\left(\frac{i}{p_j} - \frac{N_b - 1}{q_j}\right)\sin\frac{2\pi j}{m_d}. \tag{6.15}$$

6.4.3 LBM method

An alternative method to determine the solute concentrations of the coexisting phases has been developed [17] based on the Langer, Bar-on, Miller (LBM) treatment of non-linear spinodal decomposition [18]. In this model, Langer *et al.* considered a probability distribution consisting of a pair of Gaussian distributions of the form

$$P(c)=\frac{\mu_2\exp\left[\frac{-(c-\mu_1)^2}{2\sigma^2}\right]+\mu_1\exp\left[\frac{-(c-\mu_2)^2}{2\sigma^2}\right]}{(\mu_1+\mu_2)\sigma\sqrt{2\pi}}, \qquad 6.16$$

with one Gaussian distribution centered at μ_1 and the other at μ_2 and both having a width of σ. Therefore, this implementation provides an estimate of the concentrations of the coexisting phases rather than the difference in the concentration between the phases. The LBM and P_a models may be compared directly since $\mu_2 - \mu_1$ is equivalent to the $2P_a$ parameter.

In order to implement this model [17], the parameters μ_1 and μ_2 are replaced by parameters b_1 and b_2, where $b_1 = c_0 - \mu_1$ and$b_2 = c_0 - \mu_2$, i.e., the solute concentrations relative to the mean solute concentration, c_0. In addition, the pre-exponential terms are replaced by parameters a_1 and a_2, which are the relative weights of the two distributions based on the lever rule, such that $a_1 = b_2/(b_1 + b_2)$ and $a_2 = b_1/(b_1 + b_2)$.

The probability $P(i, N_b)$ is given by the following equations

$$P(i,N_b)=\frac{1}{m_d}\sum_{j=0}^{m_d}\binom{N_b}{i}p_{1j}^{i}q_{1j}^{N_b-1}\frac{4a_1}{0.9544\sqrt{2\pi}}\exp\left(\frac{-u_j^2}{2\sigma^2}\right) + \frac{1}{m_d}\sum_{j=0}^{m_d}\binom{N_b}{i}p_{2j}^{i}q_{2j}^{N_b-1}\frac{4a_2}{0.9544\sqrt{2\pi}}\exp\left(\frac{-u_j^2}{2\sigma^2}\right), \qquad 6.17$$

$$u_j=\frac{4\sigma j}{m_d}-2\sigma,$$

$$p_{1j}=c_0+b_1+u_j \quad (0<p_{1j}<1), \quad p_{2j}=c_0+b_2+u_j \quad (0<p_{2j}<1), \qquad 6.18$$

$$q_{1j}=1-p_{1j}, \quad \text{and} \quad q_{2j}=1-p_{2j}$$

Since Gaussian distributions may have significant tails, it is possible to generate solute concentrations that exceed 100% or are negative, especially in cases where the phase separation results in large concentration fluctuations.

In these cases, these unphysical concentrations are either ignored or the distribution is re-normalized while maintaining the same volume fraction.

As in the P_a method, the likelihood L is maximized with respect to b_1, b_2 and σ by maximizing eqn 6.13 and eqn. 6.17 by stepping through all combinations of possible values. Alternatively, the moment method may also be applied to find the best fit between the model and the experimental data [15,16].

6.4.4 Variation method

Another method to estimate the development of composition fluctuations is the variation or differential histogram method [19]. In this method, the parameter V is defined as the difference between the observed experimental frequency distribution, $O(i, N_b)$, and the binomial distribution, $B(i, N_b)$, calculated with the same mean concentration (i.e., eqn 6.12)

$$V = \sum_{i=0}^{N_b} \left| O(i, N_b) - B(i, N_b) \right| . \qquad 6.19$$

where N_b is the number of ions in the block. The parameter V, which is the 'distance' between these two distributions, can take values between 0 and 2. Larger values of V indicate greater concentration differences between the phases present.

The difference between the positions of the two maxima in the difference frequency distribution has also been used as a parameter to monitor the extent of decomposition. However, the precise positions of these peaks are sometimes difficult to evaluate and this parameter is not a true estimate of the composition differences between the phases. Therefore, the peak difference method should only be used to compare the atom probe results to those generated by models of the decomposition process.

6.4.5 Contingency table

The cosegregation and antisegregation behavior of the elements in a volume containing different phases may be examined with the use of contingency tables. The contingency table is created from the composition array and may be regarded as a two-dimensional frequency distribution [20,21]. The correlation between elements may be determined by comparing the numbers of composition blocks containing high or low concentrations of the two elements with those expected from a random distribution.

The general form of a two-dimensional r x c contingency table is shown in Table 6.2. In this two-part table, r is the number of rows and c is the number of columns. The first part contains the experimental observations of the numbers of blocks of data that contain specific numbers of A and B atoms and the second part contains their estimates. The observed frequency or number of observations in the ith category of the row (first variable) and the jth category of the column (second variable) is represented by n_{ij}.

The probability that an observation belongs to the ith category of the row variable and the jth category of the column variable is represented by p_{ij}. The probability that an observation in the population belongs to the ith category of the row variable without reference to the column variable is given in dot notation by $p_{i\bullet}$ and the corresponding probability of the jth category for the column variable by $p_{\bullet j}$. To test for independence of the two categories, the null hypothesis $p_{ij} = p_{i\bullet} p_{\bullet j}$ is investigated.

Table 6.2. General form of a two-dimensional r x c contingency table.

Experimental values

		Columns (Variable B)				
		1	2	•	c	total
Rows	1	n_{11}	n_{12}	•	n_{1c}	$n_{1\bullet}$
	2	n_{21}	n_{22}	•	n_{2c}	$n_{2\bullet}$
Variable	•	•	•	•	•	•
A	r	n_{r1}	n_{r2}	•	n_{rc}	$n_{r\bullet}$
	total	$n_{\bullet 1}$	$n_{\bullet 2}$	•	$n_{\bullet c}$	$n_{\bullet\bullet}$

Estimated values

		Columns (Variable B)				
		1	2	•	c	total
Rows	1	e_{11}	e_{12}	•	e_{1c}	$e_{1\bullet}$
	2	e_{21}	e_{22}	•	e_{2c}	$e_{2\bullet}$
Variable	•	•	•	•	•	•
A	r	e_{r1}	e_{r2}	•	e_{rc}	$e_{r\bullet}$
	total	$e_{\bullet 1}$	$e_{\bullet 2}$	•	$e_{\bullet c}$	$e_{\bullet\bullet}$

The estimates of the expected frequencies, e_{ij}, are given in dot notation by

$$e_{ij} = \frac{n_{i\bullet} n_{\bullet j}}{n_{\bullet\bullet}} \qquad 6.20$$

where $n_{\bullet\bullet}$ is the total number of observations and the row and column marginal totals are defined as

$$n_{i\bullet} = \sum_{j=1}^{c} n_{ij} \quad \text{and} \quad n_{\bullet j} = \sum_{i=1}^{r} n_{ij} \qquad 6.21$$

respectively. The row and column marginal totals for the estimated values are defined in a similar manner and should be identical to their experimental values. Categories with small probabilities should be combined so that the expected number of observations in a class is greater than 5.

The X^2 statistic may be evaluated from the contingency table with the following

$$X^2 = \sum_{i=1}^{r}\sum_{j=1}^{c}\frac{(n_{ij}-e_{ij})^2}{e_{ij}} . \qquad 6.22$$

If the rows and columns are independent, the X^2 statistic has approximately a χ^2 distribution with (r-1)(c-1) degrees of freedom. Combination of categories reduces the number of degrees of freedom. The test of the null hypothesis is performed by comparing the probability of the X^2 statistic with the tabulated values of the χ^2 distribution for the appropriate number of degrees of freedom (Appendix F).

The construction of contingency tables has to be performed carefully to avoid systematic errors. Suppose that the data from a specimen containing three types of atoms, A, B and C, is divided into blocks with a fixed total number of atoms. If the data exhibit a larger than average correlation between A and B atoms, then a block containing a large number of B atoms will also contain fewer than average A atoms because there are fewer than average remaining A+C atoms. To ensure that this A-B correlation artifact does not occur, the atoms should be divided into blocks that contain a constant number of C atoms.

Co-segregation or anti-segregation may be evaluated from a comparison of the experimental observations and their estimated values, as shown in Table 6.3. Co-segregation of the two elements is indicated in cases where the number of experimental observations is higher than their estimates in blocks that contain high A and high B contents. This trend is usually accompanied with blocks in which there are greater than estimated number of experimental observations with both low A and low B contents. The opposite behavior indicates anti-segregation.

6.4.6 Gibbsian interfacial excess

The amount of segregation at interfaces and grain boundaries may be quantified directly from an analysis that contains an interface with the use of a method based on the Gibbsian interfacial excess. The Gibbsian interfacial excess of element i, Γ_i, is defined [22-24] as

$$\Gamma_i = \frac{n_{ix}}{A}, \qquad 6.23$$

Table 6.3. Schematic tables showing the patterns for co-segregation and anti-segregation in contingency tables. The + indicates that the experimental value is larger than the estimated value and the - indicates the opposite.

Co-segregation

		Columns (Variable B)				
		1	2	3	4	5
Rows	1	+	+	-	-	-
	2	+	-	-	-	+
Variable	3	-	-	-	+	+
A	4	-	-	+	+	-
	5	-	+	+	-	-

Anti-segregation

		Columns (Variable B)				
		1	2	3	4	5
Rows	1	-	-	+	+	+
	2	-	+	+	+	-
Variable	3	+	+	-	-	-
A	4	+	+	-	-	-
	5	+	-	-	-	-

where n_{ix} is the excess number of solute atoms associated with the interface and A is the interfacial area over which the interfacial excess is determined. The excess number of atoms n_{ix} is given by the difference of the total number of solute atoms in the volume containing the interface, n_i, and the number of atoms of element i, $n_{i(\alpha)}$ and $n_{i(\beta)}$, in the two adjoining regions, α and β either side of the dividing surface, i.e.,

$$\Gamma_i = \frac{n_i - n_{i(\alpha)} - n_{i(\beta)}}{A}. \qquad 6.24$$

Alternatively, the Gibbsian interfacial excess may be expressed in terms of the concentrations of element i, $c_{i(\alpha)}$ and $c_{i(\beta)}$ in the α and β regions,

$$\Gamma_i = \frac{n_{\mathrm{T}}(c_i - c_{i(\alpha)}f_\alpha - c_{i(\alpha)}f_\beta)}{A} = \frac{n_{\mathrm{T}}(c_i - c_{i(\alpha)} - (c_{i(\beta)} - c_{i(\alpha)})f_\beta)}{A}, \qquad 6.25$$

where c_i is the concentration of element i in the analysis volume, n_{T} is the total number of atoms in the analysis volume, and f_α and f_β are the relative volumes of the two regions. In this model of the real system, it is assumed that the compositions of the two adjacent regions are constant up to the dividing surface and that the interface has no volume (i.e., $f_\alpha + f_\beta = 1$). If the solute concentrations are the same in the two regions (i.e., $c_{i(\beta)} = c_{i(\alpha)}$), e.g. as would be expected in the case of a grain boundary, this reduces to

$$\Gamma_i = \frac{n_{\mathrm{T}}(c_i - c_{i(\alpha)})}{A}. \qquad 6.26$$

These equations assume that the single atom detector has ideal detection efficiency. Therefore, the number of atoms detected has to be corrected for the detection efficiency.

The interfacial area depends on the curvature of the interface, the geometrical parameters of the intersection of the interface with the volume of analysis and the cross section of the single atom detector. For a planar interface intersecting the central region of a cylindrical volume of analysis, the interfacial area is given by $A = \pi r_c^2/\cos\phi$, where r_c is the local radius of the cylinder of analysis and $\phi = \cos^{-1}(\mathbf{n} \bullet \mathbf{l})$, $\mathbf{n}$ is the unit normal to the interface plane, and $\mathbf{l}$ is the unit vector parallel to the cylinder of analysis. For other cases, standard geometrical formulae for areal determinations are used.

An example of a manganese atom map containing a manganese-decorated grain boundary is shown in Fig. 6.14. The Gibbsian interfacial excess may be estimated from these data with the following method. The number of each type of atoms in the analyzed volume and the total number of atoms are determined. The detection efficiency of the single atom detector was 65%. From these data, the manganese concentration may be estimated, c_{Mn} = 4,769/804,330 = 0.00593, where n_{T} = 804,330. A small volume that encompasses the enriched region at the interface is selected and the number of atoms of each type within that volume is determined. From these data, the manganese concentration in the matrix can be estimated, $c_{\mathrm{Mn}(\alpha)}$ = (4,769-503)/(804,330-23,127) = 0.00546. In this example, the boundary was planar and parallel to the x-axis. Therefore, the area of the interface may be estimated from the product of the extents of the interface within the selected volume, A = 152.7 nm^2. The Gibbsian interfacial excess for manganese, Γ_{Mn}, is 804,330 / 0.65 * (0.00593-0.00546) / (152.7 * 10^{-18}) = 3.8 x 10^{-18} atoms m^{-2}.

The estimate of the Gibbsian interfacial excess may be compared with results from other techniques by converting to the coverage of solute on the interface. The fractional monolayer coverage of solute, Φ, may be estimated from

$$\Phi = \frac{\Gamma_i}{\Gamma_{i(\mathrm{sat})}}, \qquad 6.27$$

where $\Gamma_{i(\mathrm{sat})}$ is the number of atoms per unit area that constitutes a monolayer of coverage. However, there are a number of different interpretations as to what constitutes a monolayer of coverage [24,25]. One common definition is

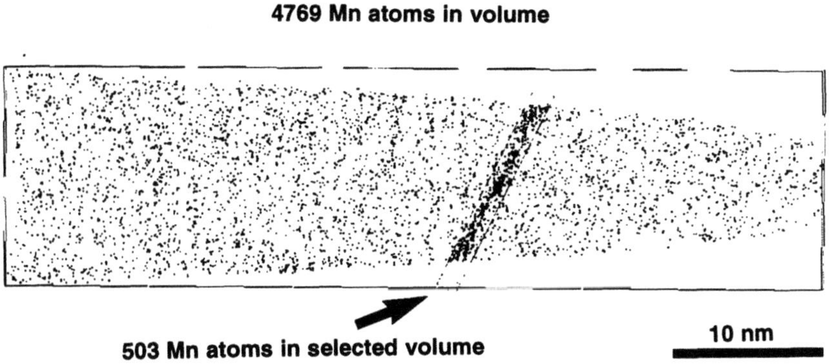

Fig. 6.14. A manganese atom map from which Gibbsian interfacial excess of manganese segregation to the grain boundary may be estimated. In the volume of analysis, there are 4,769 manganese atoms and a total of 804,330 atoms. Within the selected volume, there are 503 manganese atoms and a total of 23,127 atoms.

that one monolayer contains a^{-2} segregant atoms per unit area where a^3 is the atomic volume of the segregant [25]. In the above example, this monolayer value is $\Gamma_{Mn(sat)}$ = 1.60 x 10^{-19} atoms m^{-2} which yield a coverage at the boundary of 24% Mn. Another estimate is the number of atoms per unit area that are on the closest packed plane of the crystal structure. The saturation value of the excess may also be given by the product of the atomic density of the material, ρ_m, and the width of the segregation, w. The method used to estimate a monolayer of coverage should be stated explicitly when reporting the results.

6.5 Estimation of Dimensions

The techniques to estimate the size of features of interest in the three-dimensional data are described in this section.

6.5.1 Direct measurement

The simplest method to measure the extent of a feature is directly from one of the types of three-dimensional representations, such as the atom maps or isoconcentration surfaces. This measurement may be obtained visually from the data. In visualization software, pointers may usually be positioned at the extremes of the feature of interest in the three-dimensional representation and their coordinates displayed so that the distance between the two pointers can be determined. This measurement is generally repeated at several positions to produce an average measurement.

6.5.2 Radius of gyration and Guinier radius

The size of the feature may be estimated from the distances of the atoms from the center of the body. One common measure is the radius of gyration. The radius of gyration of an object is the radius of a hypothetical body having all of its mass concentrated at a single distance from its center of mass $(\bar{x}, \bar{y}, \bar{z})$, while having the same mass m_0 and moment of inertia I_0 as the object. This hypothetical body is a dipole in one dimension, a circle in two dimensions, and a spherical shell in three dimensions. The radius of gyration is given for the one-dimensional case, l_x, by

$$l_x = \sqrt{\frac{I_x}{m_o}} = \sqrt{\frac{\sum_{i=1}^{n} m_i (x_i - \bar{x})^2}{\sum_{i=1}^{n} m_i}}, \qquad 6.28$$

where x_i is the spatial coordinates of each atom, m_i are their masses, and n is the number of atoms in the feature. If all atoms are the same species (i.e., the same mass), this reduces to

$$l_x = \sqrt{\frac{I_x}{m_o}} = \sqrt{\frac{\sum_{i=1}^{n} (x_i - \bar{x})^2}{n}}. \qquad 6.29$$

In atom probe tomography experiments, the radius of gyration, l_g, is normally determined for the three-dimensional case from

$$l_g = \sqrt{\frac{\sum_{i=1}^{n} (x_i - \bar{x})^2 + (y_i - \bar{y})^2 + (z_i - \bar{z})^2}{n}}. \qquad 6.30$$

where x_i, y_i and z_i are the spatial coordinates of each atom.

The radius of gyration provides a parameter that is slightly smaller than the actual size of the feature. Therefore, it is common to convert the radius of gyration to an alternative parameter, the Guinier radius, r_G, which represents the actual size of the feature with the use of the following equation

$$r_G = \sqrt{\frac{5}{3}}\, l_g. \qquad 6.31$$

6.5.3 Autocorrelation function

Another statistical method that may be used to estimate both the size of a feature and the periodicity between features is the autocorrelation function (or autocorrelogram) [26]. The autocorrelation function may be expressed as

$$R_k = \frac{N_b}{N_b - k} \frac{\sum_{i=1}^{N_b+k} (c_i - c_0)(c_{i+k} - c_0)}{\sum_{i=1}^{N_b} (c_i - c_0)^2}, \qquad 6.32$$

where c_i and c_{i+k} are the concentrations of the ith and $i+k^{th}$ samples, N_b is the total number of blocks of atoms, and c_0 is the mean concentration. The parameter k is referred to as the lag. The values of the autocorrelation function R_k lie between +1 and –1. The value of the autocorrelation function at a lag of $k = 1$ is always 1, i.e., $R_1 = 1$. The range of the autocorrelation function should be restricted to lags smaller than $N_b/4$ to prevent artifacts.

In random solid solutions, the summation of the individual positive and negative correlations will generate values of the autocorrelation function that are close to zero for all lags greater than 1. In cases where there are localized regions of high solute concentration, such as from a precipitate, the values of R_k will be positive at lags smaller than the diameter of the region. Therefore the size of the region may be estimated from the first minimum in the autocorrelation function. An example of a one-dimensional autocorrelation from a random distribution of 2-nm-diameter precipitates is shown in Fig. 6.15a. Similarly, in the case of a modulated two-phase microstructure, there will be higher solute regions separated by the wavelength of the structure and a positive value of the autocorrelation function will be observed at the lag corresponding to that distance.

The standard error of the autocorrelation function is given by

$$s = \frac{1}{\sqrt{N_b - k}}. \qquad 6.33$$

For large N, the autocorrelation function is equivalent to

$$R_k = F^{-1}[\mathrm{F}(\omega)\ \mathrm{F}(\omega)^*] = F^{-1}[\mathrm{P}(\omega)], \qquad 6.34$$

i.e. the inverse Fourier transform of the power spectrum of the composition profile.

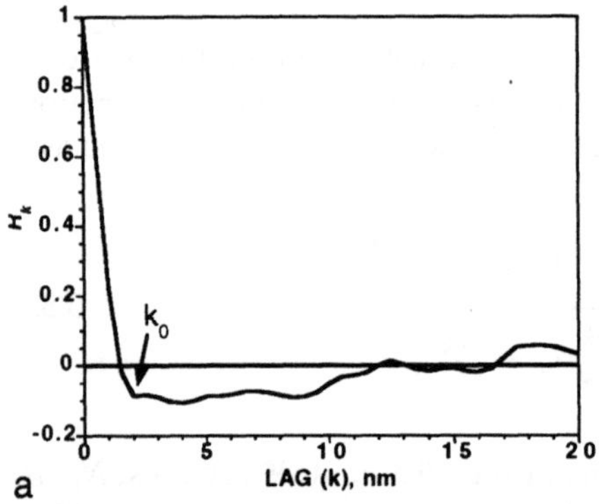

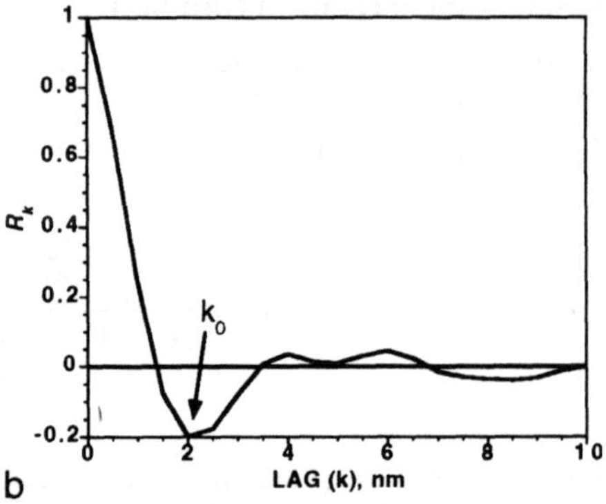

Fig. 6.15. Autocorrelation functions determined from a random distribution of 2-nm-diameter spherical precipitates. a) One-dimensional autocorrelation function and b) radial autocorrelation function. The diameter of the precipitates is indicated by the value of k_0.

This method may be extended to three dimensions to maximize the use of the three-dimensional data. The autocorrelation function in any specific direction is given by

$$R_{k_x} = \frac{1}{n_x n_y} \frac{\sum_{z=1}^{n_z}\sum_{y=1}^{n_y}\sum_{x=1}^{n_x-k_x}[c(x,y,z)-c_0(y,z)][c(x+k_x,y,z)-c_0(y,z)]}{\sum_{z=1}^{n_z}\sum_{y=1}^{n_y}\sum_{x=1}^{n_x}[c(x,y,z)-c_0(y,z)]^2}, \qquad 6.35$$

where $c(x, y, z)$ is an array of concentration values on a fixed grid such that $1 \le x \le n_x$, $1 \le y \le n_y$, $1 \le z \le n_z$ and $c_0(y,z)$ is the mean solute concentration for a line of composition blocks parallel to the x–axis. The interpretation of the three-dimensional autocorrelation function is identical to the one-dimensional variant.

An alternative variant is the radial autocorrelation function, which is given by

$$R_k = \frac{\sum_{r=0}^{r_{max}-k}(c_r - c_0)(c_{r+k} - c_0)}{\sum_{r=0}^{r_{max}}(c_r - c_0)^2}, \qquad 6.36$$

where the solute concentrations, c_r, are calculated from those atoms lying within a shell at radius r from the selected center point. The radial autocorrelation function may be applied to a plane or the entire volume. An example of a radial autocorrelation from a random distribution of 2-nm-diameter precipitates is shown in Fig. 6.15b.

6.6 Clustering and Ordering

The statistical methods for the detection of solute clustering and ordering are presented in this section. All of the methods presented in this section are derivatives of the methods developed for the classical atom probe. Ordering may also be detected by a narrowing of the frequency distribution, as discussed in §6.4.1.

6.6.1 Ladder diagrams and cumulative profiles

In ordered alloys, the compositions of the different types of planes is of interest in terms of their long range order and the site preference of solute elements. These parameters may be investigated with ladder diagrams.

Ladder diagrams are constructed by selecting a volume of analysis that is perpendicular to the ordered planes. Although this selection may in principle be performed on any data that exhibits planes, it is normal to perform the atom probe analysis such that the orientation of the specimen is optimized for the collection of atoms close to a plane that exhibits long range ordering. In an equiatomic AB alloy, the detection of each A atom is plotted as one step along the ordinate and the detection of each B alloy is plotted as one step along the abscissa. The overall slope of the resulting curve is the concentration of A in the alloy or phase. If the planes alternate as pure A and pure B atoms, as in an analysis of a B2-ordered material along an {001} direction, the ladder diagram will be composed of vertical and horizontal sections as a result of collecting atoms from the A and the B planes, etc. The concentration of solutes on each plane may be estimated from the local slope of each section in the ladder diagram. The long range order parameter, L_o, may be estimated from

$$L_o = \frac{f_\alpha - c_A}{1 - s_\alpha} = \frac{f_\beta - c_B}{1 - s_\beta} \qquad 6.37$$

where c_A and c_B are the atomic fractions of A and B atoms, s_α and s_β are the site fractions of the α and β sites in the crystal structure and f_α and f_β are the fractions of the α and β sites occupied by the correct atoms. This method may also be applied to the compositions of A and B planes that are determined by selecting volumes that are restricted individual planes, as described in §6.1.5.

Additional solute elements may be superimposed on this plot as dots or characters. The number of solute atoms in each of the different types of planes may be estimated and their site preference determined.

Ladder diagrams should be used with caution as preferential evaporation and retention are most severe in analyses near poles in ordered alloys since there are generally large composition differences between the different types of planes. It is common that the experimental parameters that normally produce reliable data do not completely eliminate preferential evaporation and retention when the analysis is performed near a pole. The reliability of the analysis can be evaluated from the following criteria: a) the overall slope of the data should yield the average composition of the alloy, b) similar numbers of atoms should be detected on the planes of different composition, and c) the spacing of the planes in an atom map should be uniform. If one or more of these criteria is not met, then the results may be misleading.

Cumulative or integral profiles are similar to ladder diagrams and only differ in that the abscissa is the total number of atoms rather than another element. The main uses of cumulative profiles are to detect segregation and cosegregation at interfaces and solute segregation in the matrix.

6.6.2 Mean separation method

The mean separation method was developed to detect the early stages of solute clustering and precipitation by examining the separation between the atoms of the element of interest [27,28]. If clustering or precipitation is present, the distribution of the separations of the atoms of interest is different from that of a random solid solution. In order to perform this type of analysis, the three-dimensional ion-by-ion data are converted into a series of square cross section columns of atoms along the analysis direction. The lateral extent of the column is kept to a minimum (typically ~0.25 nm) to preserve the spatial resolution of the position of the atoms along the column. In order to increase the length of the data and improve the statistics, the individual columns are generally appended to each other to form one long column of ions.

For a large sample, the mean separation of any distribution of solute atoms, $\hat{\mu} = (1-p)\,p$, depends only on the solute concentration, p, and is not influenced by the distribution of the solute atoms. The variance of the separations, σ^2 of the experimental separation of atoms, μ, is given by

$$\sigma^2 = \frac{\sum \mu_i^2}{N_B} - \hat{\mu}^2, \qquad 6.38$$

where N_B is the number of B atoms. The variance of the separations in the experimental column of atoms may be compared to the estimated variance of a random distribution to detect deviations from a random distribution. The estimated variance of the random solid solution is given by

$$\sigma_{\text{est}}^2 = \frac{1-p}{p^2}. \qquad 6.39$$

The standard error of the estimated variance is given by

$$\tau_{\text{est}} = \sqrt{\frac{q(1+6q+q^2)/p^4}{N_{\text{B}}}}, \qquad 6.40$$

where $q = 1 - p$. A useful parameter to compare to the other techniques in this section is $(\sigma^2 - \sigma_{\text{est}}^2)/\tau_{\text{est}}$.

6.6.3 Markov chain

The solute distribution of an element B may be investigated at the atomic scale by examining the distribution of consecutive B atoms in columns of atoms in the atom-by-atom data. These sequences of atoms would be of the form AB^nA, where n is the number of consecutive B atoms and A denotes any atom other than B. In a random solid solution, the probability $P(n)$ of detecting a chain of n consecutive B atoms in the one-dimensional sequence of atoms is given by [29]

$$P(n) = p^n q^2 \text{ or } D_{\text{exp}}(n) = N\,P(n), \qquad 6.41$$

where p is the probability of collecting a B atom, $q = 1 - p$, and N is the number of atoms in the chain. The significance of the experimental value is given by

$$s = (D_{\text{ap}}(n) - D_{\text{exp}}(n))/\sigma, \qquad 6.42$$

where $D_{\text{ap}}(n)$ is the experimentally determined number of sequences and σ may be taken as $\sqrt{Np^nq^2}$.

The tendency for the solute to order or form clusters may be investigated by comparing the number of chains experimentally observed with the number predicted for a random solid solution. The number of ABA ($n = 1$) chains experimentally observed is larger than expected in cases where the solute is in an ordered configuration. The number of ABBA, ABBBA, (i.e., $n \geq 2$) chains experimentally observed is larger than expected in cases where some of the solute is in the form of clusters. In practice, the maximum length of chain that is normally considered in this type of analysis is of the order of $n = 10$.

6.6.4 Johnson and Klotz ordering parameter

The Johnson and Klotz ordering parameter, θ, may be determined from the number of AB and BB pairs in the column of atoms [30]. As in the previous method, B is an atom of the element of interest and A is any other atom. An estimate of the ordering parameter, θ, is given by

$$\hat{\theta} = \frac{1}{2}\left[\frac{2n_{BB} + n_{AB} + m}{mp} - \frac{n_{BB} + n_{AB}}{mp^2} + \sqrt{\left(\frac{2n_{BB} + n_{AB} + m}{mp} - \frac{n_{BB} + n_{AB}}{mp^2}\right)^2 + \frac{4n_{BB}(1-2p)}{mp^3}}\right], \qquad 6.43$$

where $m = n - 1$,

$n_{AA} + n_{AB} + n_{BB} = n - 1$,

$n_{BB} = \Sigma x_i x_{i+1}$,

$n_{AB} = \Sigma[x_i(1 - x_{i+1}) + (1 - x_i)\, x_{i+1}]$,

$n_{AA} = \Sigma(1 - x_i)(1 - x_{i+1})$,

where x_i is 0 or 1 for A and B type atoms, respectively, and n is the number of atoms. If the ordering parameter θ is equal to 1, the data exhibit a random distribution of solute. If θ is less than 1, the data exhibit short range order and if θ is greater than 1, the data exhibit solute clustering. This kind of analysis is not generally appropriate for dilute alloys ($mp^2 < \sim 1$), since no n_{BB} pairs are likely to be detected.

The significance of the ordering parameter is given by $(\theta - 1) / \tau$, where τ is the standard error. The standard error is given by

$$\tau = \sqrt{\frac{\sigma^2}{n}}, \qquad 6.44$$

where the variance, σ^2, may be estimated by

$$\hat{\sigma}^2 = \frac{\hat{\theta}(1 - \hat{\theta}p)\,(1 - 2p + \hat{\theta}p^2)}{p^2\,(1 - 2p + \hat{\theta}p)}. \qquad 6.45$$

Values of the significances that are greater than 2 or less than -2 indicate non-random behavior in the distribution of solute.

The number of atoms required to evaluate the ordering parameter depends on the solute concentration, the value of the ordering parameter and the accuracy required. For solute levels greater than 10%, a minimum of 250,000 atoms is sufficient for an accuracy of 10%. This number decreases with increasing solute content.

6.7 Topological and Fractal Methods

The application of topological analysis to atom probe data is generally limited to complex microstructures such as the interconnected network structure that is observed in materials that undergo spinodal decomposition. Topological analysis is used to determine the degree of interconnectivity of the microstructure. In order to quantify the interconnectivity, the number of loops per unit volume or "handles" is determined. The handle density may be regarded as a similar parameter to the number density of isolated precipitates.

6.7.1 Percolation

There is a critical value of the volume fraction, below which a second phase tends to exist as isolated particles and above which the second phase tends to coalesce into a single interconnected region. For the latter case, a continuous path can be traced within the second phase from one surface of the volume to another. Percolation may be detected in three-dimensional atom probe by gridding the data and then assigning each voxel to a given phase based on a concentration threshold. An initial voxel in the selected phase on one surface is selected and all adjacent voxels of the same phase are joined together. Then all adjacent blocks of the same phase are joined together throughout the volume. This process is then repeated for all other voxels on the starting surface. A structure is percolated if there is at least one continuous path in the selected phase from the initial to the opposite surface of the volume. This type of analysis should be used with some caution since the volume sampled is limited and the result may not be representative of the material.

6.7.2 Skeletonization

The topological features of an object may be investigated from its skeleton. Skeletonization is also known as a medial axis transformation. One simple method to produce a skeleton is to construct the line that passes through the volume elements that mark the local midline of the object. Another method is to erode the object until only the core region remains. If the object exhibits branches or protrusions then the skeleton will also exhibit those features. The number of branches or nodes may be determined from this skeleton.

6.7.3 Handle density

The Euler characteristic may be used to determine the number of handles in a structure. The Euler number E is [31]

$$E = n_{\mathrm{n}} - n_{\mathrm{e}} + n_{\mathrm{f}} \qquad 6.46$$

where n_n is the number of nodes on the net, n_e is the number of edges, and n_f is the number of faces. The Euler number is identically two for all individual closed netted surface. For example, a simple cube may be regarded as having 8 nodes (i.e., the corners), 12 edges and 6 faces, yielding $E = 2$. This relationship does not apply if the surface has enclosed tunnels and cavities. In this case, the relationship becomes

$$E = n_n - n_e + n_f = 2 - 2n_h + 2n_c, \qquad 6.47$$

where n_h is the number of handles and n_c is the number of cavities.

6.7.4 Fractal and fracton analysis

A fractal object is one that has self–similar detail on all length scales. However, in both nature and in atom probe data, there is usually a restricted range of length scales over which scaling behavior may be observed. In other words, the microstructure is not ideally fractal.

The scaling behavior can be characterized by a fractal dimension, D, given by

$$L_m \propto s_r^D, \qquad 6.48$$

where L_m is the measured length, and s_r is the size of the ruler used to measure the length. This method may be implemented by first gridding the data, as described in §6.2.1. A concentration threshold is used to define which voxels belong to phase A, and which to phase B. The surface area of one phase, e.g., phase B, is then measured as a function of the grid spacing. Surface area can be measured from the grid in several ways, one of which is to count the number of voxels (or nodes of the composition grid) which are part of phase B but have a phase A cell as a nearest neighbor. Doubling or tripling, etc. the grid spacing can be performed without resampling the data by taking every second or third, etc. grid line in each direction. The slope of the area versus grid spacing on a log–log plot gives the fractal dimension directly.

Scaling of the microstructure may also be determined from cumulative profiles, §6.6.1. In the cumulative profile, the accumulated number of atoms of a selected element, Y(i), is plotted as a function of the total number of detected atoms [32]. This curve is a fractional Brownian motion if $Y(i) - Y(i+k)$ has a Gaussian distribution with variance

$$\sigma^2 = \langle |Y(i) - Y(i+k)|^2 \rangle = k^{2H}, \qquad 6.49$$

where the angle brackets indicate the mean value. The parameter H may therefore be determined from the slope ($2H$) of a log-log plot of the variance versus k. In a fractal Brownian motion, if H is greater than 0.5, the displacements are positively correlated, and if H is less than 0.5, the displacements are negatively correlated.

Another type of scaling may be determined from the composition data. In this method, a simulated random walk through the composition data is performed. Starting from a number of sites within the composition grid on a selected phase, the probability of returning to the starting point while remaining in that phase is measured as a function of the number of steps in the random walk (time). As the number of loops in the structure increases, and fills the volume, then the probability of returning to the starting point after a given time becomes less characteristic of a line, and more characteristic of a three dimensional object. The scaling of the probability $P(t)$ after time or number of steps n_s defines a characteristic parameter, called the fracton dimension, F, such that

$$P(t) \propto n_s^{-F/2} \qquad 6.50$$

6.8 Modelling and Simulation

There has been an increasing trend in recent years towards computer simulation and modeling [33,34]. Although simulation and modeling plays an important role in many atom probe investigations, a complete treatment of this topic is beyond the scope of this monograph.

In atom probe field ion microscopy, modeling can play an important role in understanding the microstructure and simulations can assist in the interpretation of the images and the analysis of the three-dimensional data. The converse is also true in that the atom probe results often play an important role in the validation of the models. The simulations related to atom probe field ion microscopy and atom probe tomography may be divided into three broad categories.

Thin shell and bond models have been used to simulate field ion images, as discussed in §3.2.10 and ion trajectory simulations have been used to simulate field evaporation images, as discussed in §5.5.5. Atom probe simulators have been used to generate a list of the atom positions based on a model of the microstructure and instrumental parameters such as detection efficiency and spatial resolution. These data may then be used to evaluate the effectiveness, accuracy and limitations of the statistical tools and visualization techniques. Also included in this category are the ion-optical simulations that may be used to assist in the design of the instrument.

Thermodynamic predictions of the phase compositions are routinely used to determine which phases are likely to be present after specific heat treatments together with the amount of phases. In addition, these thermodynamic predictions may be used to compare the predicted equilibrium compositions of the coexisting phases with the atom probe results.

Simulations of the microstructural development have been modeled with Monte Carlo [35,36] and molecular dynamics methods. The type of simulation performed depends on the type of microstructure under investigation. The microstructures produced by these simulations may also be subjected to the same types of statistical analyses as the atom probe data. If the fit between the simulation and the atom probe data is in reasonable agreement, the simulation may be extrapolated to either longer or shorter times or to conditions that may not be experimentally accessible.

References

1. M. K. Miller, A. Cerezo, M. G. Hetherington and G. D. W. Smith, Atom Probe Field Ion Microscopy, Oxford University Press, Oxford, UK, 1996.
2. A. Cerezo and M. G. Hetherington, *J. de Phys.*, **50-C8** (1989) 523.
3. A. Cerezo, M. G. Hetherington, J. M. Hyde, M. K. Miller, G. D. W. Smith and J. S. Underkoffler, *Surf. Sci.*, **266** (1992) 471.
4. M. K. Miller, *J. Microsc.*, **186** (1997) 1.
5. P. Pareige, Doctoral Thesis, 1994, University of Rouen, France.
6. E. O. Brigham, The Fast Fourier Transform, Prentice-Hall, Englewood Cliffs, NJ, 1974.
7. P. J. Warren, A. Cerezo and G. D. W. Smith, *Microsc. Microanal.*, **5** (1998) 89.
8. P. J. Warren, A. Cerezo and G. D. W. Smith, *Ultramicroscopy*, **73** (1998) 261.
9. P. P. Camus, D. J. Larson and T. F. Kelly, *Appl. Surf. Sci.*, **87/88** (1995) 305.
10. M. Kendall, A. Stuart and J. K. Ord, The Advanced Theory of Statistics, 4th Edition, Giffin and Box, London, UK, 1983.
11. A. Cerezo, M. G. Hetherington, J. M. Hyde and M. K. Miller. *Scripta Metall.*, **25** (1991) 1435.
12. J. M. Sassen, M. G. Hetherington, T. J. Godfrey, G. D. W. Smith, P. H. Pumphrey and K. N. Akhurst, Properties of Stainless Steel in Elevated Temperature Service, M. Prager, ed., American Society of Mechanical Engineers, New York, NY, 1987, p. 65.
13. T. J. Godfrey, M. G. Hetherington, J. M. Sassen and G. D. W. Smith, *J. de Phys.*, **49-C6** (1988) 421.
14. J. W. Cahn, *Trans. Metall. Soc. AIME*, **242** (1968) 166.

15. K. O. Bowman, M. K. Miller and L. R. Shenton, *Appl. Surf. Sci.*, **67** (1993) 424.
16. K. O. Bowman, M. K. Miller and L. R. Shenton, *Appl. Surf. Sci.*, **76/77** (1994) 403.
17. M. G. Hetherington, J. M. Hyde, M. K. Miller and G. D. W. Smith, *Surf. Sci.*, **246** (1991) 304.
18. J. S. Langer, M. Bar-on and H. D. Miller, *Phys. Rev. A*, **11** (1975) 1417.
19. D. Blavette, G. Grancher and A. Bostel, *J. de Phys.*, **49-C6** (1998) 433.
20. B. S. Everitt, The Analysis of Contingency Tables, Chapman and Hall, London, UK, 1977.
21. E. Camus and C. Abromeit, *J. Appl. Phys.*, **75** (1994) 2373.
22. J. W. Gibbs, The Collected Works of J. Willard Gibbs, Yale University Press, New Haven, CT, 1948 Volume 1.
23. B. W. Krakauer and D. N. Seidman, *Phys. Rev. B*, **48** (1993) 6724.
24. M. K. Miller and G. D. W. Smith, *Appl. Surf. Sci.*, **87/88** (1995) 243.
25. E. D. Hondros and M. P. Seah, *Int. Mater. Rev.*, **22** (1977) 262.
26. M. Kendall, A. Stuart and J. K. Ord, The Advanced Theory of Statistics, 4th Edition, Giffin and Box, London, UK, 1983, volume 3 p. 443.
27. M. G. Hetherington and M. K. Miller, *J. de Phys.*, **48-C6** (1987) 559.
28. M. G. Hetherington and M. K. Miller, *J. de Phys.*, **49-C6** (1988) 427.
29. T. T. Tsong, S. B. McLane, M. Ahmad and C. S. Wu, *J. Appl. Phys.*, **53** (1982) 4180.
30. C. A. Johnson and J. H. Klotz, *Technometrics*, **16** (1974) 483.
31. D. Hilbert and S. Cohn-Vossen, Anschauliche Geometrie, Springer-Verlag, Berlin, 1932, p. 254.
32. M. G. Hetherington and M. K. Miller, *J. de Phys.*, **50-C8** (1989) 535.
33. J. C. Russ, Computer-assisted Microscopy: The Measurement and Analysis of Images, Plenum Press, New York, 1990.
34. Computer Simulation in Materials Science, R. J. Arsenault, J. R. Beeler and D, M. Esterling, eds., ASM International, Metals Park, OH, 1988.
35. A. Cerezo, J. M. Hyde, M. K. Miller, S. C. Petts, R. P. Setna and G. D. W. Smith, *P. Roy. Soc. Lond. A*, **341** (1992) 313.
36. C. Schmuck, P. Caron, A. Hauet and D. Blavette, *Phil. Mag.*, **76A** (1997) 527.

Bibliography

This bibliography contains a comprehensive list of references to atom probe tomography applications. The references have been categorized into general reviews of the technique and by metallurgical system. Additional references to the development of the three-dimensional atom probe instrumentation are in Chapters 1 and 4 and references to data analysis are in Chapters 5 and 6. The final section in this chapter provides some sources of references to earlier classical atom probe studies.

A. Reviews

1. A. Cerezo, T. J. Godfrey and G. D. W. Smith, Application of a position-sensitive detector to atom probe microanalysis, *Rev. Sci. Instrum.*, **59** (1988) 862.
2. A. Cerezo, T. J. Godfrey and G. D. W. Smith, Development and initial applications of a position-sensitive atom probe, *J. de Phys.*, 49-C6 (1988) 25.
3. A. Cerezo, T. J. Godfrey, C. R. M. Grovenor, M. G. Hetherington, R. M. Hoyle, J. P. Jakubovics, J. A. Liddle, G. D. W. Smith and G. M. Royale, Materials analysis with a position sensitive atom probe, *J. Microsc.*, **154** (1989) 215.
4. A. Cerezo, C. R. M. Grovenor, M. G. Hetherington, B. A. Shollock and G. D. W. Smith, Characterisation of ultrafine microstructures using a position-sensitive atom probe (POSAP), Proc. Multicomponent Ultrafine Microstructures, Nov. 30-Dec. 1, 1988, Boston, MA, L. E. McCandlish, D. E. Polk, B. H. Kear and R. W. Siegel, eds., Materials Research Society, Pittsburgh, PA, **132** (1989) 151.

5. A. Cerezo, T. J. Godfrey, C. R. M. Grovenor, M. G. Hetherington, J. M. Hyde, J. A. Liddle, R. A. D. Mackenzie and G. D. W. Smith, The position sensitive atom probe: three dimensional reconstruction of atomic chemistry, *Electron Microsc. Soc. Am. Bull.,* **20** (1990) 77.
6. C. R. M. Grovenor, A. Cerezo, J. Sassen, J. A. Liddle, M. G. Hetherington, R. A. D. Mackenzie, B. A. Shollock and G. D. W. Smith, Recent developments in atom probe microanalysis of materials with ultrafine microstructures, Proc. EMAG-MICRO 89, Sept. 13-15, 1989, London, UK, P. J. Goodhew and H. Y. Elder, eds., Institute of Physics, Bristol, UK, *Inst. Phys. Conf. Ser.* No. **98** (1990) 147.
7. R. A. D. Mackenzie, G. D. W. Smith, A. Cerezo, T. J. Godfrey and J. E. Brown, Three-dimensional materials characterisation at ultrahigh resolution using a position-sensitive atom probe, Proc. 12th Intl. Congress on Electron Microscopy, Aug. 12-18, 1990, Seattle, WA, L. D. Peachey and D. B. Williams, eds., San Francisco Press, San Francisco, CA, **4** (1990) 408.
8. W. Sha, A. Cerezo, T. J. Godfrey and G. D. W. Smith, The position sensitive atom probe: a review, *J. Electron. Sci. Tech. (Chinese)*, **Suppl. 1** (1990) 17.
9. A. Cerezo, The position-sensitive atom probe, *Vacuum*, **42** (1991) 605.
10. A. Cerezo and G. D. W. Smith, The position-sensitive atom probe: reconstruction of the atomic chemistry of solids in 3-dimensions, *Mikrochim. Acta*, **II** (1991) 401.
11. W. Sha, A. Cerezo, T. J. Godfrey and G. D. W. Smith, Applications of a position-sensitive atom probe in rare metals studies, *Rare Met. (Beijing)*, **10** (1991) 248.
12. A. Cerezo, J. M. Hyde, M. K. Miller, G. Beverini, R. P. Setna, P. J. Warren and G. D. W. Smith, New dimensions in atom probe analysis, *Surf. Sci.*, **266** (1992) 481.
13. C. R. M. Grovenor, G. D. W. Smith, A. Cerezo, J. A. Liddle, R. A. D. Mackenzie, P. J. Warren, R. P. Setna, J. M. Hyde, J. E. Brown, I. Stark and B. A. Shollock, Ultra-high-resolution chemical analysis by field-ion microscopy, atom probe and position-sensitive atom-probe techniques, *Ultramicroscopy*, **47** (1992) 199.
14. M. G. Hetherington, A. Cerezo, J. M. Hyde and G. D. W. Smith, Structural analysis with the position sensitive atom probe, *Surf. Sci.,* **266** (1992) 463.
15. J. C. Riviere, Recent advances in surface analysis, *Analyst*, **117** (1992) 313.
16. G. D. W. Smith, A. Cerezo, C. R. M. Grovenor, T. J. Godfrey, and R. P. Setna, Three-dimensional reconstruction of atomic-scale composition with the position-sensitive atom probe, Proc. 50th Annual Meeting of the Electron Microscopy Society of America, Aug. 16-21, 1992, Boston, MA,

G. W. Bailey, J. Bentley, and J. A. Small, eds., San Francisco Press, San Francisco, CA, (1992) 1478.

17. D. Blavette, A. Bostel, J. M. Sarrau, B. Deconihout and A. Menand, An atom probe for 3-dimensional tomography, *Nature*, **363** (1993) 432.
18. D. Blavette, B. Deconihout, A. Bostel, J. M. Sarrau, M. Bouet and A. Menand, The tomographic atom probe - a quantitative 3-dimensional nanoanalytical instrument on an atomic scale, *Rev. Sci. Instrum.*, **64** (1993) 2911.
19. D. Blavette, B. Deconihout, A. Bostel, J. M. Sarrau and A. Menand, A new generation of 3-dimensional atom probe, *C. R. Acad. Sci. II*, **317** (1993) 1279.
20. D. Blavette, L. Letellier and B. Deconihout, Microscopy and microanalysis of interfaces on a subnanometric scale with the atom probe, *Ann. Chim. Fr.*, **18** (1993) 303.
21. D. Blavette and A. Menand, A 3-dimensional atomic analytic probe, *Recherche* **253, 24** (1993) 464.
22. D. Blavette and A. Menand, The atomic probe in physical metallurgy-present limits and future prospects, *La Revue de Métallurgie-Cahiers D Informations Techniques*, **90** (1993) 651.
23. R. A. D. Mackenzie, A. Cerezo, and G. D. W. Smith, Characterization of nanometer-scale compositional variations using field ion microscopy and the position sensitive atom probe, *Nanostruct. Mater.*, **3** (1993) 203.
24. P. Bas, S. Duval, L. Letellier, P. Pareige, B. Deconihout, S. Chambreland and D. Blavette, Atomic scale analysis of materials in three dimensions with the tomographic atom probe, Proc. 13th Intl. Congress Electron Microscopy, July 17-22, 1994, Paris, France, B. Jouffrey and C. Colliex, eds., Les Éditions de Physique, Les Ulis, France, **1** (1994) 755.
25. D. Blavette, A. Bostel and A. Menand, The tomographic atom-probe: a new nanoanalytical tool on a subnanometric scale, *J. Trace Microprobe T.*, **12** (1994) 17.
26. D. Blavette and A. Menand, New developments in atom probe techniques and potential applications to materials science, *MRS Bull.*, **19** (1994) 21.
27. A. Cerezo and G. D. W. Smith, Field-ion microscopy and atom probe analysis, *Mater. Sci. Tech. Ser.: Characterization of Materials, Part II*, E. Lifshin, ed., **2B** (1994) 514.
28. A. Cerezo, G. D. W. Smith, R. P. Setna and J. M. Hyde, Some recent advances in three-dimensional atomic-scale microanalysis with the atom probe, Proc. 13th Intl. Congress on Electron Microscopy, July 17-22, 1994, Paris, France, B. Jouffrey and C. Colliex, eds., Les Éditions de Physique, Les Ulis, France, **1** (1994) 713.

29. B. Deconihout, A. Bostel, P. Bas, S. Chambreland, L. Letellier, F. Danoix and D. Blavette, Investigation of some selected metallurgical problems with the tomographic atom probe, *Appl. Surf. Sci.*, **76/77** (1994) 145.
30. B. Deconihout, S. Chambreland and D. Blavette, The tomographic atom probe: new dimensions in materials analysis, *Adv. Mater.*, **6** (1994) 695.
31. B. Deconihout, A. Menand and D. Blavette, Recent developments of the tomographic atom probe, Proc. 52nd Annual Meeting of the Microscopy Society of America, July 31-Aug. 5, 1994, New Orleans, LA, G. W. Bailey and A. J. Garrett-Reed, eds., San Francisco Press, San Francisco, CA, (1994) 830.
32. M. Leisch, Three-dimensional field ion mass spectrometry, *Fresenius J. Anal. Chem.*, **349** (1994) 102.
33. A. Menand, A. Bostel, B. Deconihout and D. Blavette, The tomographic atom probe. A three-dimensional microanalytical instrument with subnanometer resolution, *Spectra Analyse*, **177** (1994) 33.
34. M. K. Miller, Atom probe field ion microscopy, *Vacuum*, **45** (1994) 819.
35. M. K. Miller and G. D. W. Smith, Applications of atom probe microanalysis in materials science, *MRS Bull.*, **21** (1994) 27.
36. T. Yoshimura and Y. Ishikawa, Potential and interest of nanolevel analysis using a three-dimensional atom probe, *Tetsu to Hagane*, **80** (1994) N573.
37. T. Al-Kassab, H. Wollenberger and D. Blavette, The tomographic atom probe. Chemical analysis in three dimensions by atomic resolution, *Phys. Bull.*, **51** (1995) 927.
38. P. Auger, A. Bigot, C. Schmuck, F. Danoix and D. Blavette, The role of the TAP in the investigation of some metallurgical phenomena, Proc. Intl. Conf. Advanced Materials and Technology, June 13-16, 1995, Plzen, The Czech Republic, ASM International, Materials Park, OH, **2** (1995) 40.
39. K. Hono, R. Okano and T. Sakurai, Three dimensional atom probe and its applications, *Materia (Japan)*, **34** (1995) 578.
40. D. Blavette, A. Bigot, C. Schmuck, F. Danoix and P. Auger, Three dimensional nanoanalysis with the tomographic atom-probe, *Mikrochem. Acta*, **13** (1996) 183.
41. A. Cerezo, D. Gibuoin, T. J. Godfrey, J. M. Hyde, S. Kim, R. P. Setna, S. J. Sijbrandij, F. M. Venker, J. Wilde and G. D. W. Smith, Three-dimensional atom probe microanalysis, Proc. Intl. Conf. Microstructures and Functions of Materials (ICMFM 96), Sept. 9-11, 1996, Tokyo, Japan, N. Igata, Y. Hiki, I. Yoshida and S. Sato, eds., University of Tokyo Press, Tokyo, Japan, (1996) 281.
42. A. Cerezo, D. Gibuoin, S. Kim, S. J. Sijbrandij, F. M. Venker, P. J. Warren, J. Wilde and G. D. W. Smith, Materials applications of an advanced 3-dimensional atom probe, *J. de Phys. IV*, **6-C5** (1996) 205.

43. K. Hono, Subnanometer scale microstructural characterization by a three-dimensional atom probe, *Materia (Japan)*, **35** (1996) 267.
44. M. Leisch, Tomographic atom probe: a new tool for nanoscale characterization, *Solid State Phenom.*, **47/48** (1996) 573.
45. T. Al-Kassab, H. Wollenberger and D. Blavette, Application of the tomographic atom probe to selected problems in materials science, *Z. Metallkd.*, **88** (1997) 102.
46. K. Hono and T. Sakurai, Recent atom probe studies at IMR-a comprehensive review, *Sci. Rep. RITU*, **A44** (1997) 223.
47. M. K. Miller, Three-dimensional atom probes, *J. Microsc.*, **186** (1997) 1.
48. A. Cerezo, P. J. Warren and G. D. W. Smith, The position-sensitive atom probe – a new dimension in atom probe analysis, *Microsc. Microanal.*, **4 suppl. 2** (1998) 76.
49. B. Deconihout, P. Pareige, D. Blavette, A. Bostel and A. Menand, The tomographic atom probe: a new dimension in materials analysis, *Microsc. Microanal.*, **4 suppl. 2** (1998) 78.
50. K. Hono and M. Murayama, Nanoscale microstructural analyses by atom probe field ion microscopy, *High Temp. Mater. Proc.*, **17** (1998) 69.
51. G. D. W. Smith, A. Cerezo, T. J. Godfrey, K. Seto and P. J. Warren, Studies of boundaries and interfaces on the atomic scale by atom probe FIM, Proc. D. A. Smith Symposium, Boundaries and Interfaces in Materials, Sept. 14-18, 1997, Indianapolis, IN, R. C. Pond, W. A. T. Clark, A. H. King and D. B. Williams, eds., TMS, Warrendale, PA, (1998) 131.
52. K. Hono, Atom probe microanalysis and nanoscale microstructures in metallic materials, *Acta Mater.*, **47** (1999) 3127.
53. A. Menand, E. Cadel, C. Pareige and D. Blavette, Three-dimensional atomic scale microscopy with the atom probe, *Ultramicroscopy*, **78** (1999) 63.
54. M. K. Miller, Atom probe tomography of interfaces, *Microsc. Microanal.*, **5 suppl. 2** (1999) 118.
55. M. K. Miller, Imaging elusive solute atoms, *Science*, **286** (1999) 2285.
56. M. K. Miller, The development of atom probe field ion microscopy, *Mater. Charact.*, **44** (2000) 11.

B. Phase Transformations

57. M. K. Miller, Field evaporation and field ion microscopy study of the morphology of phases produced as a result of low temperature phase transformations in the iron-chromium system, *J. de Phys.*, **50-C8** (1989) 247.
58. A. Cerezo, C. R. M. Grovenor, M. G. Hetherington, W. Sha, B. A. Shollock and G. D. W. Smith, Analysis of nanometer-sized precipitates using atom probe techniques, *Mater. Charact.*, **25** (1990) 143.

59. A. Cerezo, M. G. Hetherington, J. M. Hyde and M. K. Miller, Morphological and structural characterization of the spinodal decomposition of Fe-Cr alloys, Proc. Alloy Phase Stability and Design, April 18-20, 1990, San Francisco, CA, G. M. Stocks, D. P. Pope and A. F. Giamei, eds., Materials Research Society, Pittsburgh, PA, **186** (1991) 203.
60. A. Cerezo, J. M. Hyde, M. K. Miller, S. C. Petts, R. P. Setna and G. D. W. Smith, Atomistic modelling of diffusional phase transformations, *Phil. Trans. R. Soc. Lond. A*, **341** (1992) 313.
61. J. M. Hyde, A. Cerezo, M. G. Hetherington, M. K. Miller and G. D. W. Smith, Three-dimensional characterization and modeling of spinodally decomposed iron-chromium alloys, *Surf. Sci.*, **266** (1992) 370.
62. A. Cerezo, J. M. Hyde, M. K. Miller, R. P. Setna and G. D. W. Smith, Dynamical Ising model simulations of nucleation and growth in copper-cobalt alloys, Proc. Material Theory and Modelling, Nov. 30-Dec. 3, 1992, Boston, MA, J. Broughton, P. D. Bristowe and J. Newsam, eds., Materials Research Society, Pittsburgh, PA, **291** (1993) 623.
63. R. P. Setna, J. M. Hyde, A. Cerezo, G. D. W. Smith and M. F. Chisholm, Position sensitive atom probe study of the decomposition of a Cu-2.6 at % Co alloy, *Appl. Surf. Sci.*, **67** (1993) 368.
64. D. Blavette, F. Danoix and P. Auger, The role of AP in the investigation of phase transformations, *J. de Phys.*, **4-C3** (1994) 35.
65. J. M. Hyde, A. Cerezo, M. K. Miller and G. D. W. Smith, A critical comparison between experimental results and numerical simulations of phase-separation in the FeCr system, *Appl. Surf. Sci.*, **76/77** (1994) 233.
66. R. P. Setna, A. Cerezo, J. M. Hyde and G. D. W. Smith, Atomic scale characterization of precipitation in copper-cobalt alloys, *Appl. Surf. Sci.*, **76/77** (1994) 203.
67. D. Blavette, L. Letellier and A. Menand, Three-dimensional investigation of precipitation and interfacial segregation processes on an atomic scale, *Ann. Phys.-Paris,* **C3-20** (1995) 73.
68. F. Danoix, B. Deconihout, A. Bostel and D. Blavette, Microstructural evolution - direct observation at an atomic scale thanks to the atom probe, *J. de Phys.*, **5** (1995) 237.
69. J. M. Hyde, M. K. Miller, A. Cerezo and G. D. W. Smith, A study of the effect of ageing temperature on phase separation in Fe-45%Cr alloys, *Appl. Surf. Sci.*, **87/88** (1995) 311.
70. M. K. Miller, J. M. Hyde, A. Cerezo and G. D. W. Smith, Comparison of low temperature decomposition in Fe-Cr and duplex stainless steels, *Appl. Surf. Sci.*, **87/88** (1995) 323.
71. M. K. Miller, J. M. Hyde, M. G. Hetherington, A. Cerezo, G. D. W. Smith and C. M. Elliott, Spinodal decomposition in Fe-Cr alloys: experimental

study at the atomic level and comparison with computer models. I. Introduction and methodology, *Acta Metall. Mater.*, **43** (1995) 3385.
72. J. M. Hyde, M. K. Miller, M. G. Hetherington, A. Cerezo, G. D. W. Smith and C. M. Elliott, Spinodal decomposition in Fe-Cr alloys: experimental study at the atomic level and comparison with computer models. II. Development of domain size and composition amplitude, *Acta Metall. Mater.*, **43** (1995) 3403.
73. J. M. Hyde, M. K. Miller, M. G. Hetherington, A. Cerezo, G. D. W. Smith and C. M. Elliott, Spinodal decomposition in Fe-Cr alloys: experimental study at the atomic level and comparison with computer models. III. Development of morphology, *Acta Metall. Mater.*, **43** (1995) 3415.
74. R. Okano, K. Hono, K. Takanashi, H. Fujimori and T. Sakurai, Magnetoresistance and phase decomposition in Cr-Fe bulk alloys, *J. Appl. Phys.*, **77** (1995) 51.
75. S. Duval, S. Chambreland and B. Deconihout, Contribution of 3D atom probe to the understanding of plane-by-plane AP analyses data: application to the study of ordering in Cu_3Au, *Appl. Surf. Sci.*, **94/95** (1996) 449.
76. M. Thuvander, K. Stiller, D. Blavette and A. Menand, Grain boundary precipitation and segregation in Ni-16Cr-9Fe model materials, *Appl. Surf. Sci.*, **94/95** (1996) 343.
77. D. Blavette, C. Pareige-Schmuck, K. Stiller and F. Danoix, Atomic-scale study of transformation paths in unmixing and ordering reactions, *Ann. Phys.-Paris,* **C2-22** (1997) 29.
78. F. Soisson, C. Pareige, M. Athénes, G. Martin and D. Blavette, Kinetics of phase transformations in metallic alloys: Monte-Carlo simulations versus experiments, *Ann. Phys.-Paris,* **C2-22** (1997) 3.
79. S. Welzel, S. Duval, E. Camus and D. Blavette, Calculation of the long-range order parameter from atom probe data, *Intermetallics*, **5** (1997) 609.
80. D. Blavette, B. Deconihout, S. Chambreland and A. Bostel, Three-dimensional imaging of chemical order with the tomographic atom-probe, *Ultramicroscopy*, **70** (1998) 115.
81. S. Duval, S. Chambreland, A. Loiseau and D. Blavette, Investigation of ordering kinetics in Cu(3)Au with the tomographic atom probe, *J. Mater. Res.*, **13** (1998) 1502.
82. C. Pareige-Schmuck, F. Soisson and D. Blavette, Ordering and phase separation in low supersaturated Ni-Cr-Al alloys: 3D atom probe and Monte Carlo simulation, *Mater. Sci. Eng.*, **A250** (1998) 99.
83. I. Rozdilsky, A. Cerezo, G. D. W. Smith and A. Watson, Atomic scale study of precipitate/matrix interfaces in a metallic alloy, Proc. Phase Transformations and Systems Driven far from Equilibrium, Dec. 1-5, 1997, Boston, MA, E. Ma, P. Bellon, M. Atzmon and R. Trivedi, eds., Materials Research Society, Pittsburgh, PA, **481** (1998) 521.

84. C. Pareige, F. Soisson, G. Martin and D. Blavette, Ordering and phase separation in Ni-Cr-Al: Monte Carlo simulations vs three-dimensional atom probe, *Acta Mater.*, **47** (1999) 1889.
85. K. Hono, D. H. Ping, M. Ohnuma and H. Onodera, Cu clustering and Si partitioning in the early crystallization stage of an $Fe_{73.5}Si_{13.5}B_9Nb_3Cu_1$ amorphous alloy, *Acta Mater.*, **47** (1999) 997.
86. K. Hono and D. H. Ping, Atom probe studies of nanocrystallization of amorphous alloys, Proc. Intl. Conf. Solid-Solid Phase Transformations PTM '99, May 24-28, 1999, Kyoto, Japan, M. Koiwa, K. Otsuka and T. Miyazaki, eds., The Japan Institute of Metals, Sendai, Japan, (1999) 1207.

C. Steels

87. W. Sha, A. Cerezo, T. J. Godfrey and G. D. W. Smith, Atom probe studies of early stages of precipitation reactions in maraging steels. 1. Co-containing and Ti-containing C-300 steel, *Scripta Metall. Mater.*, **26** (1992) 517.
88. W. Sha, A. Cerezo, T. J. Godfrey and G. D. W. Smith, Atom probe studies of early stages of precipitation reactions in maraging steels. 2. Ti-free model alloy and Co-free T-300 steel, *Scripta Metall. Mater.*, **26** (1992) 523.
89. W. Sha, G. D. W. Smith and A. Cerezo, Atom probe field-ion microscopy study of aging behaviour of a model Fe-Ni-Co-Mo maraging steel, *Surf. Sci.*, **266** (1992) 378.
90. F. Danoix, P. Auger, S. Chambreland and D. Blavette, A 3-D study of G-phase precipitation in spinodally decomposed α ferrite by tomographic atom-probe analysis, *Microsc. Microanal. M.*, **5** (1994) 121.
91. Y. Ishikawa and T. Yoshimura, Electrochemical and atom probe studies of cast and aged duplex stainless steel, *Mater. T. JIM*, **35** (1994) 895.
92. S. Ohkido, Y. Ishikawa, and T. Yoshimura, POSAP analysis of the oxide alloy interface in stainless steel, *Appl. Surf. Sci.*, **76** (1994) 261.
93. T. Yoshimura, Y. Ishikawa and S. Ohkido, Analysis of oxide film on stainless steel via position-sensitive atom-probe, *J. Vac. Sci Technol.*, **12A** (1994) 2544.
94. P. Auger, P. Pareige, M. Akamatsu and D. Blavette, APFIM investigation of clustering in neutron-irradiated Fe-Cu alloys and pressure vessel steels, *J. Nucl. Mater.*, **225** (1995) 225.
95. P. Auger, F. Danoix, O. Grisot, J. P. Massoud and J. C. Van Duysen, Spinodal decomposition in duplex stainless steels, investigated by atom probe, neutron scattering and thermoelectric measurements, *Ann. Phys.-Paris*, **C3-20** (1995) 143.
96. M. Leisch and E. Kozeschnik, Position-sensitive atom probe study of precipitates in high speed steel, *Vacuum*, **46** (1995) 1155.

97. P. Pareige, P. Auger, P. Bas and D. Blavette, Direct observation of copper precipitation in a neutron irradiated FeCu alloy by 3D atomic tomography, *Scripta Metall. Mater.*, **33** (1995) 1033.
98. P. Pareige, S. Welzel and P. Auger, Clustering effects under irradiation in Fe-0.1% Cu alloy: an atomic scale investigation with the tomographic atom probe, *J. de Phys. IV*, **6-C5** (1996) 229.
99. H. G. Read and K. Hono, Phase-separation, partitioning and precipitation in MA956, an ODS ferritic stainless steel, *J. de Phys. IV*, **6-C5** (1996) 223.
100. K. Stiller, F. Danoix and A. Bostel, Investigation of precipitation in a new maraging stainless steel, *Appl. Surf. Sci.*, **94/95** (1996) 326.
101. P. Pareige, P. Auger, S. Miloudi, J. C. Van Duysen and M. Akamatsu, Microstructural evaluation of the CHOOZ A PWR surveillance program material : small angle neutron scattering and tomographic atom probe studies, *Ann. Phys.-Paris*, **C2-22** (1997) 117.
102. H. G. Read, H. Murakami and K. Hono, Al partitioning in MA-956, an ODS ferritic stainless steel, *Scripta Mater.*, **36** (1997) 355.
103. H. G. Read, W. T. Reynolds, K. Hono and T. Tarui, APFIM and TEM studies of drawn pearlitic wire, *Scripta Mater.*, **37** (1997) 1221.
104. F. Danoix, D. Julien, X. Sauvage and J. Copreaux, Direct evidence of cementite dissolution in drawn pearlitic steels observed by tomographic atom probe, *Mater. Sci. Eng.*, **A250** (1998) 8.
105. F. Danoix, M. Hedin, P. Auger, F. Cortial and A. Buchon, Evidence of pre-precipitation stages during cooling of X2CrNiMoCuN 25.06.03 superduplex steel, *Mater. Sci. Eng.*, **A250** (1998) 14.
106. M. K. Miller, K. F. Russell, P. Pareige, M. J. Starink and R. C. Thomson, Low temperature copper solubilities in Fe-Cu-Ni, *Mater. Sci. Eng.*, **A250** (1998) 49.
107. M. Murayama, Y. Katayama and K. Hono, Phase separation and precipitation in a PH 17-4 stainless steel by prolonged aging at 400°C, *Microsc. Microanal.*, **4 suppl. 2** (1998) 96.
108. K. Stiller, F. Danoix and M. Hättestrand, Mo precipitation in a 12Cr-9Ni-4Mo-2Cu maraging steel, *Mater. Sci. Eng.*, **A250** (1998) 22.
109. K. Stiller, M. Hättestrand and F. Danoix, Precipitation in 9Ni-12Cr-2Cu maraging steels, *Acta Mater.*, **46** (1998) 6063.
110. G. D. W. Smith, A. Cerezo, T. J. Godfrey, J. Wilde and F. M. Venker, Atom probe studies of the nanochemistry of steels, *Microsc. Microanal.*, **4 suppl. 2** (1998) 104.
111. M. H. Hong, W. T. Reynolds, T. Tarui and K. Hono, Atom probe and transmisssion electron microscopy investigations of heavily drawn pearlitic steel wire, *Metall. Mater. Trans. A*, **30** (1999) 717.

112. T. F. Kelly, D. J. Larson, M. K. Miller and J. E. Flinn, 3D mapping of light element segregation in 316SS by atom probe, *Microsc. Microanal.*, **5 suppl. 2** (1999) 116.
113. T. F. Kelly, D. J. Larson, M. K. Miller and J. E. Flinn, Three-dimensional atom probe investigation of vanadium nitride precipitates and the role of oxygen and boron in rapidly-solidified 316 stainless steel, *Mater. Sci. Eng.*, **A270** (1999) 19.
114. E. A. Kenik, J. T. Busby, M. K. Miller, A. M. Thuvander and G. S. Was, Origin and influence of pre-existing segregation on radiation-induced segregation in austenitic stainless steels, Proc. Microstructural Processes in Irradiated Materials, Nov. 30-Dec. 4, 1998, Boston, MA, S. J. Zinkle, G. E. Lucas, R. C. Ewing and J. S. Williams, eds., Materials Research Society, Pittsburgh, PA, **540** (1999) 445.
115. E. A. Kenik, J. T. Busby, M. K. Miller, A. M. Thuvander and G. Was, Grain boundary segregation and irradiation-assisted stress corrosion cracking of stainless steel, *Microsc. Microanal.*, **5 suppl. 2** (1999) 760.
116. M. Murayama, K. Hono, H. Hiurkawa, T. Ohmura and S. Matsuoka, The combined effect of molybdenum and nitrogen on the fatigued microstructure of 316 type austenitic stainless steel, *Scripta Mater.*, **41** (1999) 467.
117. M. Murayama, Y. Katayama and K. Hono, Microstructural evolution in a 17-4 PH stainless steel after aging at 400°C, *Metall. Mater. Trans. A*, **30** (1999) 345.
118. K. Seto, D. J. Larson, P. J. Warren and G. D. W. Smith, Grain boundary segregation in boron added interstitial free steels studied by 3-dimensional atom probe, *Scripta Mater.*, **40** (1999) 1029..
119. F. Danoix and P. Auger, Atom probe studies of the Fe-Cr System and stainless steels aged at intermediate temperatures: A review, *Mater. Charact.*, **44** (2000) 177.
120. M. K. Miller, P. Pareige and M. G. Burke, Understanding pressure vessel steels: An atom probe perspective, *Mater. Charact.*, **44** (2000) 235.
121. P. Hofer, M. K. Miller, S. S. Babu, S. A. David and H. Cerjak, Atom probe field ion microscopy investigation of boron containing martensitic 9% chromium steel, *Metall. Trans. A*, in press.
122. P. Pareige, P. Auger, S. Welzel, J. -C. Van Duysen and S. Miloudi, Annealing of a low copper steel: hardness, SANS, atom probe and thermoelectric power investigations, Proc. Effects of Radiation on Materials: 19th Intl. Symp., ASTM STP 1366, June 16-18, 1998, Seattle, WA, M. L. Hamilton, A. S. Kumar, S. T. Rosinski and M. L. Grossbeck, eds., American Society for Testing and Materials, West Conshohocken, PA, 1999, in press.

123. J. Wilde, A. Cerezo and G. D. W. Smith, Three-dimensional atomic-scale mapping of a Cottrell atmosphere around a dislocation in iron, *Scripta Mater.*, in press.

D. Superalloys

124. L. Letellier, A. Bostel and D. Blavette, Direct observation of boron segregation at grain boundaries in Astroloy by 3D atomic tomography, *Scripta Metall. Mater.*, **30** (1994) 1503.
125. L. Letellier, M. Guttmann and D. Blavette, Atomic scale investigation of grain boundary microchemistry in the nickel based superalloy Astroloy with a 3-dimensional atom probe, *Phil. Mag. Lett.*, **70** (1994) 189.
126. D. Blavette, S. Chambreland, S. Duval, L. Letellier and A. Menand, Observation and analysis at subnanometer resolution of single crystal superalloys by field ion microscopy, Proc. Coll. Superalliages Monocristallins, Toulouse-Seilh, France, B. Clemont and A. Coujou, eds., Soc. Nat. de Construction de Moteurs d'Avion, Toulouse, France, (1995).
127. H. Murakami, P. J. Warren and H. Harada, Atom probe microanalyses of some Ni-base single crystal superalloys, Proc. 3rd Intl. Charles Parsons Turbine Conf., April 25-27, 1995, Newcastle-upon-Tyne, UK, R. D. Conroy, M. J. Goulette and A. Strang, eds., Institute of Materials, London, UK, **1** (1995) 343.
128. D. Blavette, P. Duval, L. Letellier and M. Guttmann, Atomic-scale APFIM and TEM investigation of grain boundary microchemistry in Astroloy nickel base superalloys, *Acta Mater.*, **44** (1996) 4995.
129. D. Blavette, L. Letellier, P. Duval and M. Guttmann, Atomic-scale investigation of grain boundary segregation in Astroloy with a three dimensional atom-probe, *Mater. Sci. Forum*, **207-209** (1996) 79.
130. C. Schmuck, F. Danoix, P. Caron, A. Hauet and D. Blavette, Atomic scale investigation of ordering and precipitation processes in a model NiCrAl alloy, *Appl. Surf. Sci.*, **94/95** (1996) 273.
131. T. Yoshimura, Y. Ishikawa, M. Saito, K. Hidaka and T. Ohashi, Atomic-scale analysis of microchemical changes in a Ni-based superalloy single crystal during creep, *Mater. Sci. Res. Int.*, **2** (1996) 13.
132. C. Schmuck, P. Caron, A. Hauet and D. Blavette, Ordering and precipitation of γ' phase in low supersaturated Ni- Cr-Al model alloy: an atomic scale investigation, *Phil. Mag.*, **76A** (1997) 527.
133. D. Blavette, The role of atom probe in the study of nickel base superalloys, *Microsc. Microanal.*, **4 suppl. 2** (1998) 106.
134. P. J. Warren, A. Cerezo and G. D. W. Smith, An atom probe study of the distribution of rhenium in a nickel-based superalloy, *Mater. Sci. Eng.*, **A250** (1998) 88.

135. D. Blavette, F. Soisson, G. Martin and C. Pareige, Ordering and phase separation in model superalloys: 3D atom probe versus Monte Carlo simulation, Proc. Intl. Conf. Solid-Solid Phase Transformations PTM '99, May 24-28, 1999, Kyoto, Japan, M. Koiwa, K. Otsuka and T. Miyazaki, eds., The Japan Institute of Metals, Sendai, Japan, (1999) 65.
136. M. K. Miller, Atom probe studies of phase transformations in nickel-based superalloys, Proc. Intl. Conf. Solid-Solid Phase Transformations PTM'99, May 24-28, 1999, Kyoto, Japan, M. Koiwa, K. Otsuka and T. Miyazaki, eds., The Japan Institute of Metals, Sendai, Japan, (1999) 73.
137. M. K. Miller, S. S. Babu and M. G. Burke, Intragranular precipitation in Alloy 718, *Mater. Sci. Eng.*, **A270** (1999) 14.
138. D. Blavette, E. Cadel and B. Deconihout, The role of the atom probe in the study of nickel base superalloys, *Mater. Charact.*, **44** (2000) 133.

E. Intermetallics

139. A. Menand, A. Huguet and A. Nerac-Partaix, Interstitial solubility in γ-phase and α_2-phase of TiAl-based alloys, *Acta Mater.*, **44** (1996) 4729.
140. A. Nerac-Partaix and A. Menand, Atom probe analysis of oxygen in ternary TiAl alloys, *Scripta Mater.*, **35** (1996) 199.
141. S. Kim and G. D. W. Smith, AP-FIM investigation on γ-based titanium aluminides, *Mater. Sci. Eng.*, **A239/240** (1997) 229.
142. H. Murakami, P. J. Warren, T. Kumeta, Y. Koizumi and H. Harada, Microstructural evolution of $\beta 2$ (NiTi-based) / β'(Ni_2TiAl-based) two phase quaternary alloys, *J. Surf. Analysis*, **3** (1997) 516.
143. P. Warren, Y. Murakami, Y. Koizumi and H. Harada, Phase separation in NiTi-Ni_2TiAl alloy system, *Mater. Sci. Eng.*, **A223** (1997) 17.
144. S. Kim, G. D. W. Smith, S. G. Roberts and A. Cerezo, Alloying elements characterisation in gamma-based titanium aluminides by APFIM, *Mater. Sci. Eng.*, **A250** (1998) 77.
145. A. Menand, H. Zapolsky-Tatarenko and A. Nerac-Partaix, Atom-probe investigations of TiAl alloys, *Mater. Sci. Eng.*, **A250** (1998) 55.
146. H. Wood, G. D. W. Smith and A. Cerezo, Short range order and phase separation in Ti-Al alloys, *Mater. Sci. Eng.*, **A250** (1998) 83.
147. D. Blavette, E. Cadel, A. Fraczkiewicz, and A Menand, Three-dimensional atomic scale imaging of impurity segregation to line defects, *Science*, **286** (1999) 2317.
148. E. Cadel, D. Lemarchand, A.–S. Gay, A. Fracziewicz and D. Blavette, Atomic scale investigation of boron nanosegregation in FeAl intermetallics, *Scripta Mater.*, **41** (1999) 421.
149. D. J. Larson, C. T. Liu and M. K. Miller, The alloying effects of tantalum on the microstructure of an α_2+γ titanium aluminide, *Mater. Sci. Eng.*, **A270** (1999) 1.

150. D. J. Larson and M. K. Miller, Solute partitioning and interfacial segregation in TiAl-based alloys, Proc. High Temperature Ordered Intermetallic Alloys VIII, Nov. 30-Dec. 4, 1998, Boston, MA, E. P. George, M. Yamaguchi and M. J. Mills, eds. Materials Research Society, Pittsburgh, PA, **552** (1999) KK2.9.1.
151. H. Liew, G. D. W. Smith, A. Cerezo and D. J. Larson, Experimental studies of the phase separation mechanism in Ti-15at%Al, *Mater. Sci. Eng.*, **A270** (1999) 9.
152. S. Kim, D. J. Larson and G. D. W. Smith, Alloying elements partitioning in TiAl-Ru intermetallic alloys, *Intermetallics*, **7** (1999) 1283.
153. D. J. Larson and M. K. Miller, Atom probe field ion microscopy characterization of nickel and titanium aluminides, *Mater. Charact.*, **44** (2000) 159.

F. Aluminum Alloys

154. C. R. M. Grovenor, B. A. Shollock and J. M. Knowles, Position sensitive atom probe studies of the composition of Ω and θ′ precipitates in Al-Cu-Mg-Ag alloys, *J. de Phys.*, **50-C8** (1989) 377.
155. P. J. Warren, C. R. M. Grovenor and J. S. Crompton, Field-ion microscope/atom-probe analysis of the effect of RRA heat treatment on the matrix strengthening precipitates in alloy Al-7150, *Surf. Sci.*, **266** (1992) 342.
156. K. Hono, Y. Zhang, A. P. Tsai, A. Inoue and T. Sakurai, Solute partitioning in partially crystallized Al-Ni-Ce(-Cu) metallic glasses, *Scripta Metall. Mater.*, **32** (1995) 191.
157. C. Schmuck, P. Auger, F. Danoix and D. Blavette, Quantitative analysis of GP zones formed at room temperature in a 7150 Al-based alloy, *Appl. Surf. Sci.*, **87/88** (1995) 228.
158. P. J. Warren and C. R. M. Grovenor, Comparison to STEM and atom probe methods for chemical analysis of grain boundaries in commercial Al alloys, *Mater. Sci. Forum*, **189/190** (1995) 115.
159. A. Bigot, F. Danoix, P. Auger, D. Blavette and A. Menand, 3D reconstruction and analysis of GP zones in Al-1.7 at % Cu: a tomographic atom probe investigation, *Appl. Surf. Sci.*, **94/95** (1996) 261.
160. A. Bigot, F. Danoix, P. Auger, D. Blavette and A. Reeves, Tomographic atom probe study of age hardening precipitation in industrial AlZnMgCu (7050) alloy, *Mater. Sci. Forum*, **217-222** (1996) 695.
161. A. Bigot, P. Auger, S. Chambreland, D. Blavette and A. Reeves, Atomic scale imaging and analysis of T′ precipitates in Al-Mg-Zn alloys, *Microsc. Microanal. M.*, **8** (1997) 103.
162. E. S. Humphreys, P. J. Warren and A. Cerezo, Characterisation of a rapidly solidified Al-V-Fe alloy, *Mater. Sci. Eng.*, **A250** (1998) 158.

163. M. Murayama and K. Hono, Three dimensional atom probe analysis of pre-precipitate clustering in an Al-Cu-Mg-Ag alloy, *Scripta Mater.,* **38** (1998) 1315.
164. M. Murayama and K. Hono, Atom probe characterization of preprecipitate clusters and precipitates in Al-Mg-Si(-Cu) alloys, Proc. 6th Intl. Conf. On Aluminum Alloys, July 5-10, 1998, Toyohashi, Japan, T. Sato, S. Kumai, T. Kobayashi and Y. Murakami, eds., The Japan Institute of Light Metals, Tokyo, Japan **2** (1998) 837.
165. M. Murayama, K. Hono, M. Saga and M. Kikuchi, Atom probe studies on the early stages of precipitation in Al-Mg-Si, *Mater. Sci. Eng.,* **A250** (1998) 127.
166. M. Murayama, L. Reich and K. Hono, Clustering and segregation of Ag and Mg atoms in the nucleation and growth stage of Ω and T_1 precipitates in Al-Cu(-Li) alloys, *Microsc. Microanal.*, **4 suppl. 2** (1998) 116.
167. L. Reich, M. Murayama, and K. Hono, 3DAP study of the effect of Mg and Ag additions on precipitation in Al-Cu(-Li) alloys, Proc. 6th Intl. Conf. On Aluminum Alloys, July 5-10, 1998, Toyohashi, Japan, T. Sato, S. Kumai, T. Kobayashi and Y. Murakami, eds., The Japan Institute of Light Metals, Tokyo, Japan **2** (1998) 645.
168. L. Reich, M. Murayama and K. Hono, Evolution of Ω phase in an Al-Cu-Mg-Ag alloy - a three-dimensional atom probe study, *Acta Mater.,* **46** (1998) 6053.
169. K. Hono, M. Murayama and L. Reich, Clustering and segregation of Mg and Ag atoms in the precipitation processes in Al(-Li)-Cu-Mg-Ag alloys, Proc. Intl. Conf. Solid-Solid Phase Transformations PTM '99, May 24-28, 1999, Kyoto, Japan, M. Koiwa, K. Otsuka and T. Miyazaki, eds., The Japan Institute of Metals, Sendai, Japan, (1999) 165.
170. E. S. Humphreys, P. J. Warren, A. Cerezo and G. D. W. Smith, Microstructural and chemical analysis of nanoscale particles in rapidly solidified Al-V-Fe, *Mater. Sci. Eng.,* **A 270** (1999) 48.
171. C. R. Hutchinson, K. Raviprasad and S. P. Ringer, The effects of Si and Ag on precipitation in Al-Cu-Mg alloys, Proc. Intl. Conf. Solid-Solid Phase Transformations PTM '99, May 24-28, 1999, Kyoto, Japan, M. Koiwa, K. Otsuka and T. Miyazaki, eds., The Japan Institute of Metals, Sendai, Japan, (1999) 169.
172. S. K. Maloney, K. Hono, I. J. Polmear and S. P. Ringer, The chemistry of precipitates in an aged Al-2.1 Zn-1.7 Mg at % alloy, *Scripta Mater.,* **41** (1999) 1031.
173. M. Murayama and K. Hono, Pre-precipitate clusters and precipitation processes in Al-Mg-Si alloys, *Acta Mater.,* **47** (1999) 1537.
174. M. Murayama and K. Hono, The effect of aging on the clustering and precipitation process in Al-Mg-Si alloys, Proc. Intl. Conf. Solid-Solid

Phase Transformations PTM '99, May 24-28, 1999, Kyoto, Japan, M. Koiwa, K. Otsuka and T. Miyazaki, eds., The Japan Institute of Metals, Sendai, Japan, (1999) 165..

175. K. Stiller, P. J. Warren, V. Hansen, J. Angenete and J. Gjonnes, Investigation of precipitation in an Al-Zn-Mg alloy after two step ageing treatment at 100° and 150°C, *Mater. Sci. Eng.*, **A270** (1999) 55.
176. S. P. Ringer and K. Hono, Microstructural evolution and age hardening in aluminium alloys: atom probe field ion microscopy and transmission electron microscopy studies, *Mater. Charact.*, **44** (2000) 101.
177. L. Reich, S. P. Ringer and K. Hono, Origin of the initial rapid age-hardening in an Al-1.7 Mg-1.1 Cu alloy, *Phil. Mag. Lett.*, **79** (1999) 639.

G. Multilayers and Films

178. A. Cerezo, T. J. Godfrey, C. R. M. Grovenor and G. D. W. Smith, Surface analysis with a position-sensitive atom probe, *J. Phys. Condens. Matter*, **1** (1989) SB99.
179. A. Cerezo, J. A. Liddle and C. R. M. Grovenor, Position-sensitive atom probe analysis of multiple quantum-well structures, Proc. Microscopy of Semiconducting Materials, April 10-13, 1989, Oxford, UK, A. G. Cullis and J. L. Hutchinson, eds., Institute of Physics, Bristol, UK, *Inst. Phys. Conf. Ser.* No. **100** (1989) 75.
180. A. Cerezo, J. A. Liddle, C. R. M. Grovenor, A. G. Norman and G. D. W. Smith, Application of a position sensitive atom probe to the analysis of the chemistry and morphology of multi-quantum well interfaces, Proc. Characterization of the Structure and Chemistry of Defects in Materials, Nov. 28-Dec. 3, 1989, Boston, MA, C. Bennett, C. Larson, M. Ruhle and D. N. Seidman, eds., Materials Research Society, Pittsburgh, PA, **138** (1989) 335.
181. J. A. Liddle, A. G. Norman, A. Cerezo and C. R. M. Grovenor, Application of position sensitive atom probe to the study of the microchemistry and morphology of quantum well interfaces, *Appl. Phys. Lett.*, **54** (1989) 1555.
182. R. A. D. Mackenzie, G. D. W. Smith, A. Cerezo, J. A. Liddle, C. R. M. Grovenor and M. G. Hetherington, Characterisation of multilayer materials using a position-sensitive atom probe, Proc. 12th Intl. Congress on Electron Microscopy, Aug. 12-18, 1990, Seattle, WA, L. D. Peachey and D. B. Williams, eds., San Francisco Press, San Francisco, CA, **4** (1990) 624.
183. A. Cerezo, J. M. Hyde, M. G. Hetherington and A. K. Petford-Long, Atom probe studies of interfaces in metallic multilayers, Proc. Phase Transformation Kinetics in Thin Films, April 29-May 1, 1991, Anaheim, CA, M. Chen, M. Thompson, R. Schwarz, and M. Libera, eds., Materials Research Society, Pittsburgh, PA, **230** (1991) 79.

184. R. A. D. Mackenzie, A. Cerezo and C. R. M. Grovenor, 3-dimensional visualization of semiconductor multi-quantum-well interfaces, Proc. Microscopy of Semiconducting Materials, March 25-28, 1991, Oxford, UK, A. G. Cullis and N. J. Long, eds., Institute of Physics, Bristol, UK, *Inst. Phys. Conf. Ser.* No. **117** (1991) 87.
185. A. Cerezo, C. R. M. Grovenor, and R. Ozsanlav, Atomic scale chemistry of Co- and Ni-Si(100) interfaces, Proc. Conf., March 25-28, 1991, Oxford, UK, A. G. Cullis and N. J. Long, eds., Institute of Physics, Bristol, UK, *Inst. Phys. Conf. Ser.* No. **117** (1991) 97.
186. A. K. Petford-Long, A. Cerezo and J. M. Hyde, Atom probe analysis and modelling of interfaces in magnetic multilayers, *Ultramicroscopy*, **47** (1992) 367.
187. A. K. Petford-Long, R. C. Doole, A. Cerezo, J. S. Conyers and J. P. Jakubovics, The effects of annealing on magnetic domain structure and interface profile in sputtered Fe/Cr multilayer films, *J. Magn. Magn. Mater.*, **126** (1993) 117.
188. S. Ohkido, Y. Ishikawa and T. Yoshimura, POSAP analysis of the oxide alloy interface in stainless steel, *Appl. Surf. Sci.*, **76/77** (1994) 261.
189. W. Athenstaedt and M. Leisch, A three-dimensional atom-probe study of preferential sputtering of a Ni91Pt9 alloy, *Appl. Surf. Sci.*, **87/88** (1995) 318.
190. W. Athenstaedt and M. Leisch, Three-dimensional atom probe study of the segregation behavior of Ni91Pt9 alloy after sputtering, *Fresenius J. Anal. Chem.*, **353** (1995) 753.
191. K. Hono, K. Yeh, Y. Maeda and T. Sakurai, Three-dimensional atom probe analysis of a sputter-deposited Co-Cr thin film, *Appl. Phys. Lett.*, **66** (1995) 58.
192. G. D. W. Smith, A. Cerezo and S. Poulston, Atom probe studies of catalyst surfaces: structure and chemistry at atomic level, Proc. Microscopy and Microanalysis '95, Aug. 13-17, 1995, Kansas City, MO, G. W. Bailey, M. H. Ellisman, R. A. Hennigar and N. J. Zaluzec, eds., Jones and Begell Publishing, New York, NY, (1995) 420.
193. T. Al-Kassab, M.-P. Macht, V. Naundorf, H. Wollenberger, S. Chambreland, F. Danoix and D. Blavette, Characterization of sputter-deposited multilayers of Ni and Zr with APFIM/TAP, *Appl. Surf. Sci.*, **94/95** (1996) 306.
194. W. Athenstaedt and M. Leisch, The segregation behaviour of a Pt90Rh10 alloy studied with a three-dimensional atom-probe, *Appl. Surf. Sci.*, **94/95** (1996) 403.
195. A. M. Baker, A. Cerezo and A. K. Petford-Long, Interfacial diffusion studies in Co-Pd layered films, *J. Magn. Magn. Mater.*, **156** (1996) 83.

196. J. Nishimaki, K. Hono, N. Hasegawa and T. Sakurai, Three dimensional atom probe analysis of Co-Cr-Ta thin film, *Appl. Phys. Lett.*, **69** (1996) 3095.
197. W. Tieber, W. Athenstaedt and M. Leisch, 3D-atom probe study of oxygen adsorption on stepped platinum surfaces, *Fresenius J. Anal. Chem.*, **358** (1997) 116.
198. D. J. Larson, A. K. Petford-Long, A. Cerezo, G. D. W. Smith, D. T. Foord and T. C. Anthony, Three-dimensional atom probe field-ion microscopy observation of Cu/Co multilayer film structures, *Appl. Phys. Lett.*, **73** (1998) 1125.
199. D. J. Larson, A. K. Petford-Long, A. Cerezo, T. C. Anthony and M. K. Miller, Atom probe field ion microscopy of multilayer thin films, *Microsc. Microanal.*, **5 suppl. 2** (1999) 150.

H. Miscellaneous Studies

200. W. Sha and F. Zhu, A position-sensitive atom probe approach to characterization of Al_2O_3 dispersion strengthened Cu alloy, *J. Nucl. Mater.*, **186** (1992) 288.
201. K. Hono and T. Sakurai, Atom probe studies of nanostructured alloys, *Appl. Surf. Sci.*, **87/88** (1995) 166.
202. M. Leisch, Position sensitive atom probe study of segregation in molybdenum alloys, *Fresenius J. Anal. Chem.*, **353** (1995) 251.
203. A. Eckschlager, W. Athenstaedt and M. Leisch, Atom probe study of the segregation behaviour of molybdenum containing alloys, *Vacuum*, **48** (1997) 617.
204. M. C. Dolezal, W. Athenstaedt and M. Leisch, Atom probe study of the segregation behaviour of molybdenum-rhenium alloys, *Vacuum*, **50** (1998) 323.
205. A. Eckschlager, W. Athenstaedt and M. Leisch, Atom probe study on the segregation behaviour of the binary alloy Pt95Mo5, *Fresenius J. Anal. Chem.*, **371** (1998) 672.
206. K. Hono, D. H. Ping and M. Ohnuma, APFIM and HREM studies of nanocomposite soft and hard magnetic materials, *Microsc. Microanal.*, **4 suppl. 2** (1998) 108.
207. D. H. Ping, K. Hono and S. Hirosawa, Partitioning of Ga and Co atoms in a $Fe_3B/Nd_2Fe_{14}B$ nanocomposite magnet, *J. Appl. Phys.*, **83** (1998) 7769.
208. D. Blavette, F. Danoix, E. Cadel, G. Geandier and A. Menand, Contributions of the tomographic atom probe to the observation and analysis of grain boundaries, *J. de Phys. IV*, **9** (4) (1999) 113.
209. M. W. Chen, A. Inoue, T. Sakurai, D. H. Ping and K. Hono, Impurity oxygen redistribution in a nanocrystallized $Zr_{65}Cu_{15}Al_{10}Pd_{10}$ metallic glass, *Appl. Phys. Lett.*, **74** (1999) 812.

210. K. Hono and D. H. Ping, APFIM studies of nanocomposite soft and hard magnetic materials, *Mater. Sci. Forum*, **307** (1999) 69.
211. M. K. Miller, R. B. Schwarz and Yi He, Decomposition in $Pd_{40}Ni_{40}P_{20}$ bulk metallic glass, Proc. Bulk Metallic Glasses, Nov. 30-Dec. 4, 1998, Boston, MA, W. L. Johnson, C. T. Liu and A. Inoue, eds., Materials Research Society, Pittsburgh, PA, **554** (1999) 9.
212. D. H. Ping, K. Hono and A. Inoue, Oxygen distribution in Zr-based metallic glasses, Proc. Bulk Metallic Glasses, Nov. 30-Dec. 4, 1998, Boston, MA, W. L. Johnson, C. T. Liu and A. Inoue, eds., Materials Research Society, Pittsburgh, PA, **554** (1999) 3.
213. D. H. Ping, K. Hono and A. Inoue, Oxygen distribution in $Zr_{65}Cu_{15}Al_{10}Pd_{10}$ nanocrystalline alloys, *Mater. Sci. Forum*, **307** (1999) 31.
214. D. H. Ping, K. Hono, H. Kanekiyo and S. Hirosawa, Mechanism of grain size refinement of $Fe_3B/Nd_2Fe_{14}B$ nanocomposite permanent magnet by Cu addition, *J. Appl. Phys.*, **85** (1999) 2448.
215. D. H. Ping, K. Hono, H. Kanekiyo and S. Hirosawa, Effect of Cu addition on the microstructure and magnetic properties of an $Fe_3B/Nd_2Fe_{14}B$ nanocomposite magnet, *J. Mag. Soc. Japan*, **23** (1999) 1101.
216. M. Thuvander, M. K. Miller and K. Stiller, Segregation and precipitation during heat treatment at 600°C in a model Alloy 600, *Mater. Sci. Eng.*, **A270** (1999) 38.
217. K. Hono and D. H. Ping, Atom probe studies of nanocrystallization of amorphous alloys, *Mater. Charact.*, **44** (2000) 203.
218. D. H. Ping, K. Hono, H. Kanekiyo and S. Hirosawa, Microstructural evolution of $Fe_3B/Nd_2Fe_{14}B$ nanocomposite magnets microalloyed with Cu and Nb, *Acta Mater.*, in press.
219. D. H. Ping, K. Hono, H. Kanekiyo and S. Hirosawa, Microalloying effect of Cu and Nb on the microstructure and magnetic properties of $Fe_3B/Nd_2Fe_{14}B$ nanocomposite permanent magnets, *IEEE T. Magn.*, **35** (1999) 3262.
220. Y. Q. Wu, D. H. Ping and K. Hono, Atom probe characterization of an α-$Fe/Nd_2Fe_{14}B$ nanocomposite magnet with a remaining amorphous phase, *IEEE T. Magn.*, **35** (1999) 3295.

I. Other Sources of References to Atom Probe Studies

221. M. K. Miller, A. Cerezo, M. G. Hetherington and G. D. W. Smith, Atom Probe Field Ion Microscopy, Oxford University Press, Oxford, UK, 1996.
222. R. E. Thurstans and J. M. Walls, Field-Ion Microscopy and Related Techniques: A Bibliography, 1951 - 1978, Warwick Publishing Co., Birmingham, UK, (1980).

223. Atom Probe Field Ion Microscopy and Related Techniques: A Bibliography, Oak Ridge National Laboratory Technical Reports: ORNL-TM 11157 (1988), ORNL-TM 11370 (1989), ORNL-TM 11696 (1990), ORNL-TM 12005 (1991), ORNL-TM 12223 (1992), ORNL-TM 12625 (1993), and ORNL-TM 12926 (1994).
224. Proc. 27th International Field Emission Symposium, Tokyo, Japan, 1980, Y. Yashiro and N. Igata, eds., University of Tokyo, Tokyo, Japan, (1980).
225. Proc. 29th International Field Emission Symposium, Gothenburg, Sweden, 1982, H.-O. Andren and H. Norden, eds., Almqvist and Wiksell, Stockholm, Sweden, (1982).
226. Proc. 31st International Field Emission Symposium, Paris, France, 1984, P. Sudraud and P. Ballongue, eds., *J. de Phys.*, **45-C9** (1984).
227. Proc. 32nd International Field Emission Symposium, Wheeling, WV, 1985, M. K. Miller and S. S. Brenner, eds., *J. de Phys.*, **47-C2** (1986).
228. Proc. 33rd International Field Emission Symposium, Berlin, Germany, 1986, J. H. Block, W. A. Schmidt and M. K. Miller, eds., *J. de Phys.*, **47-C7** (1986).
229. Proc. 34th International Field Emission Symposium, Osaka, Japan, 1987, S. Nakamura, O. Nishikawa and M. K. Miller, eds., *J. de Phys.*, **48-C6** (1987).
230. Proc. 35th International Field Emission Symposium, Oak Ridge, TN, 1988, M. K. Miller, ed., *J. de Phys.*, **49-C6** (1988).
231. Proc. 36th International Field Emission Symposium, Oxford, UK, 1989, A. Cerezo, M. K. Miller and G. D. W. Smith, eds., *J. de Phys.*, **50-C8** (1989).
232. Proc. 37th International Field Emission Symposium, Albuquerque, NM, 1990, G. L. Kellogg, J. A. Panitz and P. R. Schwoebel, eds., *Surf. Sci.*, **246** (1991).
233. Proc. 38th International Field Emission Symposium, Vienna, Austria, 1991, J. Mitterauer, ed., *Surf. Sci.*, **266** (1992).
234. Proc. 39th International Field Emission Symposium, Halifax, Canada, 1992, M. K. Miller and H. J. Kreuzer, eds., *Appl. Surf. Sci.*, **67** (1993).
235. Proc. 40th International Field Emission Symposium, Nagoya, Japan, 1993, M. Yamamoto, F. Okuama and M. K. Miller, eds., *Appl. Surf. Sci.*, **76/77** (1994).
236. Proc. 41st International Field Emission Symposium, Rouen, France, 1994, D. Blavette, A. Menand and M. K. Miller, eds., *Appl. Surf. Sci.*, **87/88** (1995).
237. Proc. 42nd International Field Emission Symposium, Madison, WI, 1995, M. K. Miller and T. F. Kelly, eds., *Appl. Surf. Sci.*, **94/95** (1996).
238. Proc. 43rd International Field Emission Symposium, Moscow, Russia, 1996, M. K. Miller, A. L. Survorov and R. Z. Bahktizin, eds., *J. de Phys. IV, coll. C5, Suppl. J. de Phys. III*, **6** (1996).

239. Proc. 44th International Field Emission Symposium, Tsukuba, Japan, 1997, K. Hono and M. Tsukada, eds., *Ultramicroscopy,* **73** (1998) and *Mater. Sci. Eng.*, **A250** (1998).
240. Proc. 45th International Field Emission Symposium, Irbid, Jordan, 1998, M. Mousa and F. Danoix, eds., *Ultramicroscopy*, **79** (1999) and *Mater. Sci. Eng.*, **A270** (1999).

See also §1.2

Appendix A

INTERPLANAR SPACINGS

Cubic: $$\frac{1}{d^2} = \frac{h^2 + k^2 + l^2}{a^2}$$

Tetragonal: $$\frac{1}{d^2} = \frac{h^2 + k^2}{a^2} + \frac{l^2}{c^2}$$

Hexagonal: $$\frac{1}{d^2} = \frac{4}{3}\left(\frac{h^2 + hk + k^2}{a^2}\right) + \frac{l^2}{c^2}$$

Rhombohedral:

$$\frac{1}{d^2} = \frac{\left(h^2 + k^2 + l^2\right)\sin^2\alpha + 2(hk + kl + hl)\left(\cos^2\alpha - \cos\alpha\right)}{a^2\left(1 - 3\cos^2\alpha + 2\cos^3\alpha\right)}$$

Orthorhombic: $$\frac{1}{d^2} = \frac{h^2}{a^2} + \frac{k^2}{b^2} + \frac{l^2}{c^2}$$

where d is the interplanar spacing of the hkl plane and a, b, c and α are the unit cell parameters.

VOLUMES OF UNIT CELLS

Cubic: $V = a^3$

Tetragonal: $V = a^2c$

Hexagonal: $V = \frac{\sqrt{3}}{2}a^2c$

Rhombohedral: $V = a^3\sqrt{1 - 3\cos^2\alpha + 2\cos^3\alpha}$

Orthorhombic: $V = abc$

ANGLES BETWEEN CRYSTAL PLANES

The angle, α, between $h_1k_1l_1$ and $h_2k_2l_2$ planes in a crystal is given by:

Cubic:
$$\alpha = \cos^{-1}\left\{\frac{h_1h_2 + k_1k_2 + l_1l_2}{\sqrt{\left(h_1^2 + k_1^2 + l_1^2\right)\left(h_2^2 + k_2^2 + l_2^2\right)}}\right\}$$

Tetragonal:
$$\alpha = \cos^{-1}\left\{\frac{\dfrac{h_1h_2 + k_1k_2}{a^2} + \dfrac{l_1l_2}{c^2}}{\sqrt{\left(\dfrac{h_1^2 + k_1^2}{a^2} + \dfrac{l_1^2}{c^2}\right)\left(\dfrac{h_2^2 + k_2^2}{a^2} + \dfrac{l_2^2}{c^2}\right)}}\right\}$$

Hexagonal:
$$\alpha = \cos^{-1}\left\{\frac{h_1h_2 + k_1k_2 + \frac{1}{2}\left(h_1k_2 + h_2k_1\right) + \frac{3a^2}{4c^2}l_1l_2}{\sqrt{\left(h_1^2 + k_1^2 + h_1k_1 + \frac{3a^2}{4c^2}l_1^2\right)\left(h_2^2 + k_2^2 + h_2k_2 + \frac{3a^2}{4c^2}l_2^2\right)}}\right\}$$

Rhombohedral:
$$\alpha = \cos^{-1}\left\{\begin{matrix}\dfrac{a^4d_1d_2}{V^2}\left(\sin^2\alpha(h_1h_2 + k_1k_2 + l_1l_2)\right) + \\ (\cos^2\alpha - \cos\alpha)(k_1l_2 + k_2l_1 + l_1h_2 + l_2h_1 + h_1k_2 + h_2k_1)\end{matrix}\right\}$$

Orthorhombic:
$$\alpha = \cos^{-1}\left\{\frac{\dfrac{h_1h_2}{a^2} + \dfrac{k_1k_2}{b^2} + \dfrac{l_1l_2}{c^2}}{\sqrt{\left(\dfrac{h_1^2}{a^2} + \dfrac{k_1^2}{b^2} + \dfrac{l_1^2}{c^2}\right)\left(\dfrac{h_2^2}{a^2} + \dfrac{k_2^2}{b^2} + \dfrac{l_2^2}{c^2}\right)}}\right\}$$

where V is the volume of the unit cell and d_1 and d_2 are the interplanar spacings of the $h_1k_1l_1$ and $h_2k_2l_2$ planes, respectively.

ANGLES BETWEEN PLANES IN CUBIC CRYSTAL

	$\{h_2k_2l_2\}$						
$\{h_1k_1l_1\}$	**100**	**110**	**111**	**210**	**211**	**221**	**310**
100	0 90						
110	45 90	0 60 90					
111	54.7	35.3 90	0 70.5				
210	26.6 63.4 90	18.4 50.8 71.6	39.2 75.0	0 36.9 53.1 66.4 78.5 90			
211	35.3 65.9	30 54.7 73.2 90	19.5 61.9 90	24.1 43.1 56.8 79.5 90	0 33.6 48.2 60 70.5 80.4		
221	48.2 70.5	19.5 45 76.4 90	15.8 54.7 78.9	26.6 41.8 53.4 63.4 72.7 90	17.7 35.3 47.1 65.9 74.2 82.2	0 27.3 38.9 63.6 83.6 90	
310	18.4 71.6 90	26.6 47.9 63.4 77.1	43.1 68.6	8.1 32.0 45 64.9 73.6 81.9	25.4 49.8 58.9 75 82.6	32.5 42.5 58.2 65.1 84.0	0 25.8 36.9 53.1 72.5 84.3
311	25.2 72.5	31.5 64.8 90	29.5 58.5 80.0	19.3 47.6 66.1 82.3	10.0 42.4 60.5 75.8 90	25.2 45.3 59.8 72.5 84.2	17.6 40.3 55.1 67.6 79.0 90
320	33.7 56.3 90	11.3 54.0 66.9 78.7	36.8 80.8	7.1 29.7 41.9 60.3 68.2 75.6 82.9	25.1 37.6 55.5 63.1 83.5	22.4 42.3 49.7 68.3 79.3 84.7	15.3 37.9 52.1 58.3 74.7 79.9
321	36.7 57.7 74.5	19.1 40.9 55.5 67.8 79.1	22.2 51.9 72.0 90	17.0 33.2 53.3 61.4 69.0 83.1 90	10.9 29.2 40.2 49.1 56.9 70.9 77.4 83.7 90	11.5 27.0 36.7 57.7 63.6 74.5 79.7 84.9	21.6 32.3 40.5 47.5 53.7 59.5 65.0 75.3 85.2 90
331	46.5 76.7	13.3 49.5 71.1 90	22.0 48.5 82.4	22.6 44.1 59.1 72.1 84.1	20.5 41.5 68.0 79.2	6.2 32.7 57.6 67.5 85.6	29.5 43.5 54.5 64.2 90

Appendix B

USEFUL CONSTANTS AND CONVERSIONS

Quantity	Symbol	Value	Units
atomic mass unit	u	1.6605402	10^{-27} kg
Avogadro's constant	N_A	6.0221367	10^{23} mol^{-1}
Boltzmann's constant	k_B	1.380658	10^{-23} J K^{-1}
electron mass	m_e	9.1093897	10^{-31} kg
electron volt	eV	1.60217733	10^{-19} J
elementary charge	e	1.60217733	10^{-19} C
molar gas constant	R	8.314510	Jmol^{-1}K^{-1}
Planck's constant	h	6.626176	10^{-34} Js

$\pi = 3.1415926536$
e = 2.7182818285

1 mbar = 10^2 Pa
1 torr = 1.33322×10^2 Pa

INTERCONVERSION OF ATOMIC AND WEIGHT PERCENTS

Atomic percent of element i, $A_i = \dfrac{100\dfrac{W_i}{M_i}}{\sum_{j=1}^{n}\dfrac{W_j}{M_j}}$.

Weight percent of element i, $W_i = \dfrac{A_i\, M_i}{\sum_{j=1}^{n} A_j\, M_j}$,

where M_i is the atomic weight of element i, and n is the number of elements.

Appendix C

PREDICTIONS OF LOW TEMPERATURE EVAPORATION FIELDS AND CHARGE STATES FOR THE ELEMENTS

Evaporation fields and charge states calculated from the simple image hump model by T. T. Tsong, *Surf. Sci.*, **70** (1978) 211 (reproduced with permission). Additional information from H. N. Southworth and B. Ralph, *J. Microscopy*, **90** (1969) 167.

Λ = Sublimation Energy
I_1, I_2 and I_3 = 1st, 2nd and 3rd Ionization Potentials
ϕ = Work Function
F_1, F_2 and F_3 = Calculated evaporation fields for singly, doubly and triply charged ions. Expected states (i.e., lowest values) are underlined.
F_{obs} = experimental value

Elem.	Λ	I_1	I_2	I_3	ϕ	F_1	F_2	F_3	F_{obs}
			(eV)				(V nm^{-1})		
Li	1.65	5.392	76.638	122.451	2.5	14	520	1000	
Be	3.33	9.322	18.211	153.893	3.9	53	46	770	34
B	5.92	8.30	25.15	37.9	4.6	64	79	103	
C	7.40	11.26	24.38	47.89	4.34	142	103	155	
Na	1.13	5.139	47.286	71.64	2.3	11	210	360	
Mg	1.53	7.646	15.035	80.143	3.7	21	25	220	
Al	3.34	5.986	18.828	28.447	4.1	19	35	50	
Si	4.67	8.15	16.34	33.49	4.8	45	33	60	
K	0.941	4.34	31.63	45.72	2.2	7	87	150	
Ca	1.825	6.113	11.871	50.91	2.7	19	18	100	
Ti	4.855	6.82	13.58	27.49	4.0	41	26	43	
V	5.30	6.74	14.65	29.31	4.1	44	30	49	
Cr	4.10	6.766	16.50	30.96	4.6	27	29	51	
Mn	2.98	7.435	15.640	33.67	3.8	30	30	60	
Fe	4.29	7.90	16.16	30.651	4.4	42	33	54	35
Co	4.387	7.86	17.06	33.50	4.4	43	37	63	36
Ni	4.435	7.635	18.168	35.17	5.0	35	36	65	35
Cu	3.50	7.726	20.292	36.83	4.6	30	43	77	30
Zn	1.35	9.394	17.964	39.72	3.8	33	39	84	
Ga	2.78	5.999	20.51	30.71	4.1	15	39	56	
Ge	3.98	7.88	15.93	34.22	4.8	35	29	58	
As	3.0	9.81	18.633	28.351	4.7	46	42	54	

Elem.	Λ	I_1	I_2	I_3	ϕ	F_1	F_2	F_3	F_{obs}
			(eV)				(V nm^{-1})		
Rb	0.858	4.177	27.28	40	2.1	6	69	110	
Y	4.4	6.38	12.24	20.5	3.1	40	23.5		
Zr	6.316	6.84	13.13	22.99	4.2	56	28	35	
Nb	7.47	6.88	14.32	25.04	4.0	74	37	45	35
Mo	6.810	7.10	16.15	27.16	4.2	65	41	51	46
Ru	6.615	7.37	16.76	28.47	4.5	62	41	54	
Rh	5.752	7.46	18.08	31.06	4.8	49	41	59	46
Pd	3.936	8.34	19.43	32.92	5.0	37	41	63	
Ag	2.96	7.576	21.49	34.83	4.6	24	45	72	
Cd	1.160	8.993	16.908	37.48	4.1	25	31	70	
In	2.6	5.786	18.869	28.03	4.1	12	31	46	
Sn	3.12	7.344	14.632	30.502	4.4	26	23	46	
Sb	2.7	8.641	16.53	25.3	4.6	32	30	40	
Cs	0.827	3.894	25.1	35	2.1	5	55	85	
Ba	1.86	5.212	10.004		2.5	15	13		
La	4.491	5.577	11.06	19.175	3.3	32	18	24	
Hf	6.35	7.0	14.9	23.3	3.5	67	39	43	
Ta	8.089	7.89	16	22	4.2	96	48	44	
W	8.66	7.98	18	24	4.5	102	57	52	57
Re	8.10	7.88	17	26	5.1	82	45	49	48
Os	(7)	8.7	17	25	4.6	86	48	50	
Ir	6.93	9.1	17	27	5.3	80	44	50	53
Pt	5.852	9.0	18.56	28	5.3	63	45	53	48
Au	3.78	9.225	20.5	30	4.3	53	54	66	35
Hg	0.694	10.437	18.756	34.2	4.5	31	38	66	
Tl	1.87	6.108	20.428	29.83	3.7	13	38	57	
Pb	2.04	7.416	15.032	31.937	4.1	20	23	52	
Bi	2.15	7.289	16.69	25.56	4.3	18	27	39	
U	4	6			3.3	31			

Appendix D

STEREOGRAPHIC PROJECTIONS

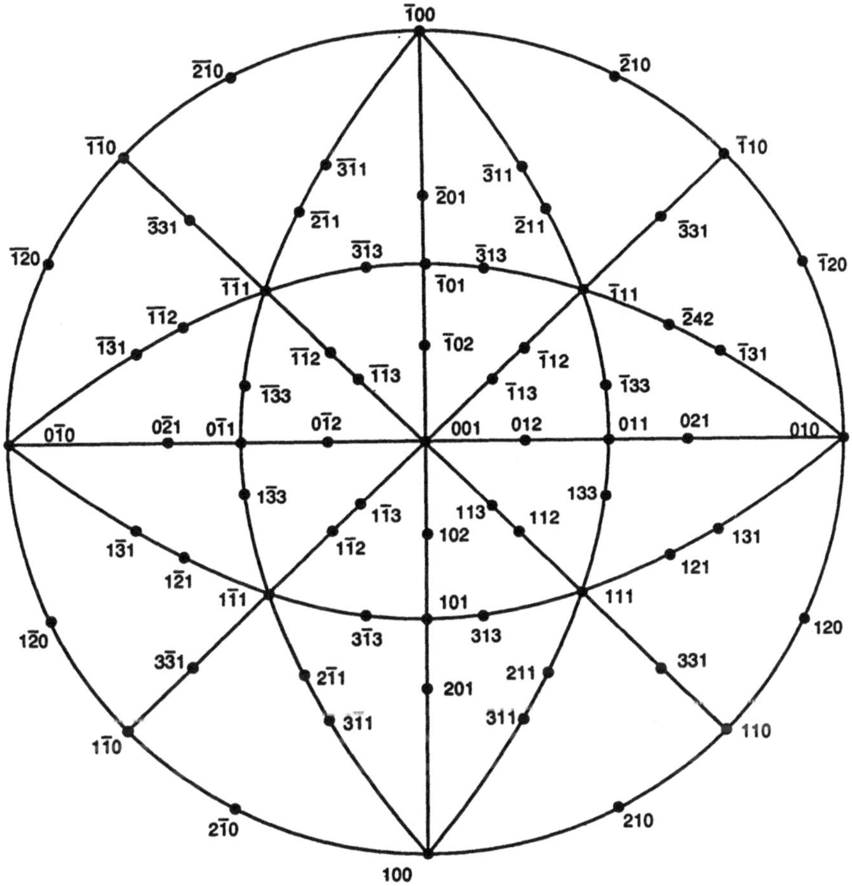

Fig. D1. Stereographic projection of a cubic lattice centered on the (001) plane (4-fold symmetry).

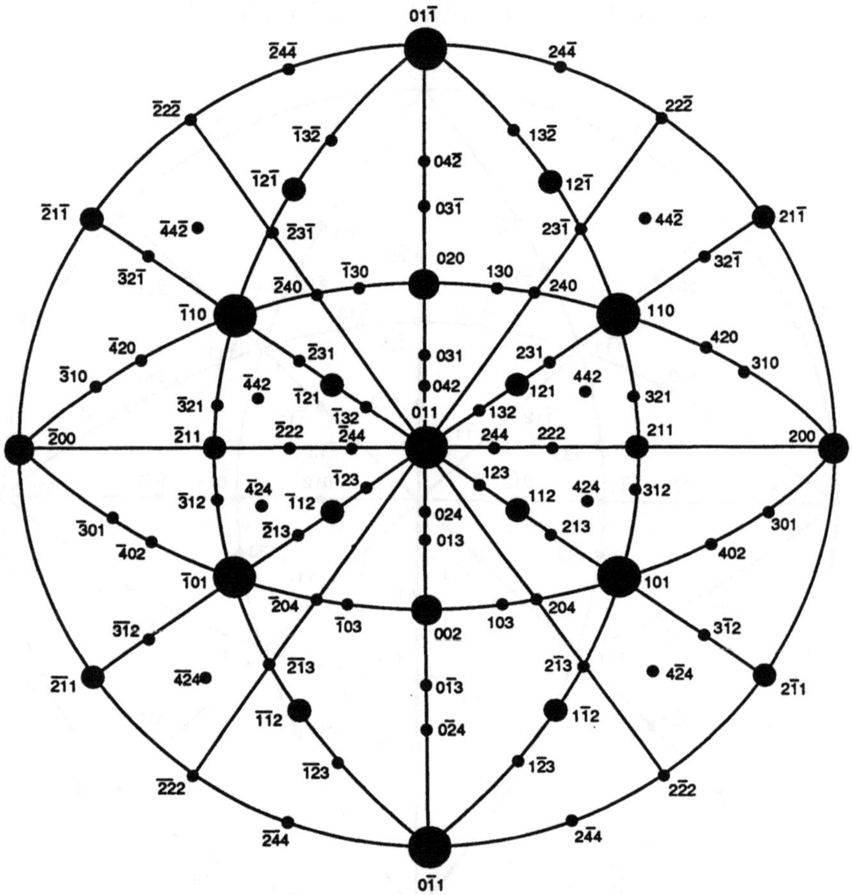

Fig. D2. Stereographic projection of a body centered cubic lattice centered on the (011) plane (2-fold symmetry). The size of the dots on the poles indicates the relative prominence of the planes.

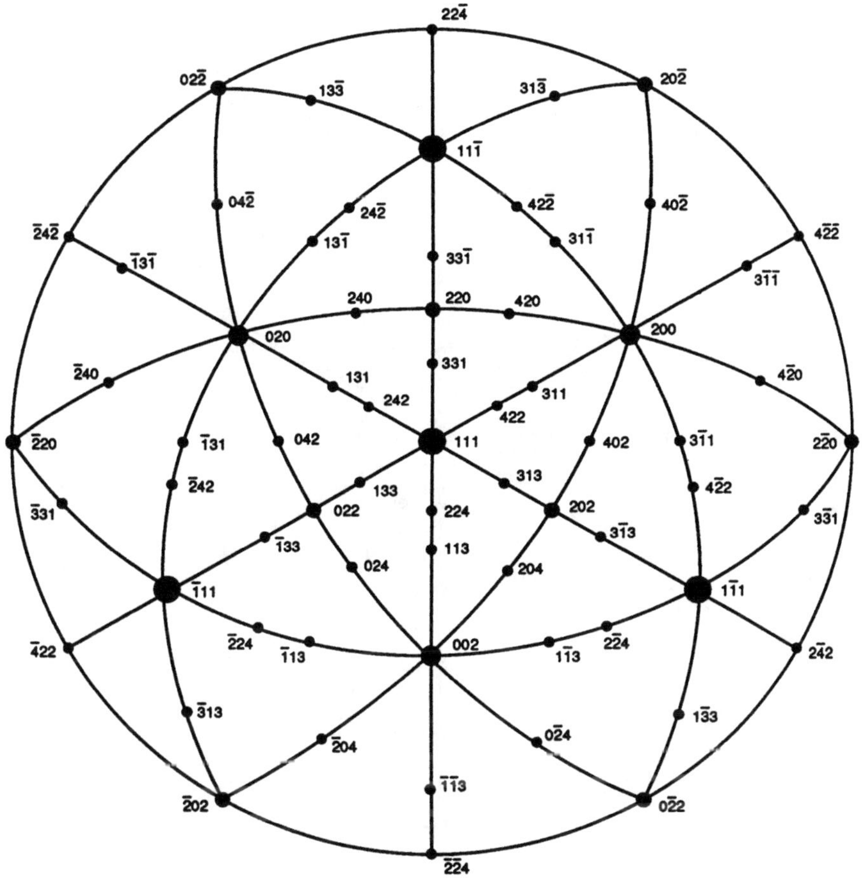

Fig. D3. Stereographic projection of a face centered cubic lattice centered on the (111) plane (3-fold symmetry). The size of the dots on the poles indicates the relative prominence of the planes.

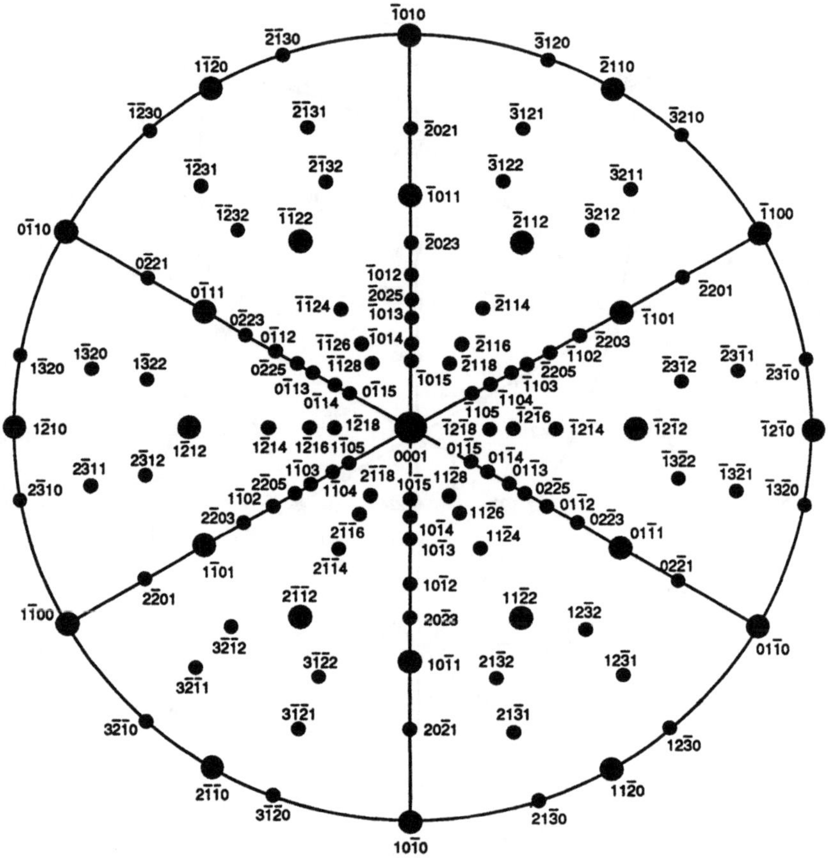

Fig. D4. Stereographic projection of a hexagonal close packed lattice with a c/a ratio of 1.633 centered on the (0001) plane (6-fold symmetry). The size of the dots on the poles indicates the relative prominence of the planes.

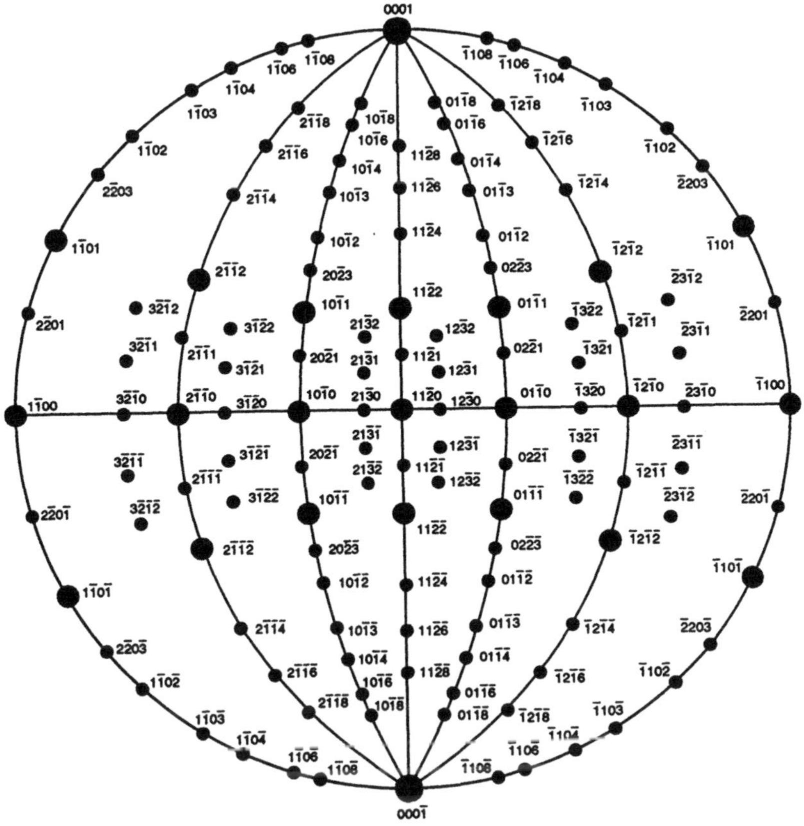

Fig. D5. Stereographic projection of a hexagonal close packed lattice with a c/a ratio of 1.633 centered on the (11$\bar{2}$0) plane (2-fold symmetry). The size of the dots on the poles indicates the relative prominence of the planes.

Appendix E

PERCENTAGE POINTS OF THE χ^2 DISTRUBUTION

Probability of a deviation greater than χ^2

Degrees of freedom	0.99	0.95	0.90	0.80	0.70	0.50	0.30	0.20	0.10	0.05	0.01	0.001
1	$.0^3157$	$.0^2393$	0.0158	0.0642	0.148	0.455	1.074	1.642	2.706	3.841	6.635	10.828
2	0.0201	0.103	0.211	0.446	0.713	1.386	2.408	3.219	4.605	5.991	9.210	13.816
3	0.115	0.352	0.584	1.005	1.424	2.366	3.665	4.642	6.251	7.815	11.345	16.266
4	0.297	0.711	1.064	1.649	2.195	3.357	4.878	5.989	7.779	9.488	13.277	18.467
5	0.554	1.145	1.610	2.343	3.000	4.351	6.064	7.289	9.236	11.070	15.086	20.515
6	0.872	1.635	2.204	3.070	3.828	5.348	7.231	8.558	10.645	12.592	16.812	22.458
7	1.239	2.167	2.833	3.822	4.671	6.346	8.383	9.803	12.017	14.067	18.475	24.322
8	1.646	2.733	3.490	4.594	5.527	7.344	9.524	11.030	13.362	15.507	20.090	26.125
9	2.088	3.325	4.168	5.380	6.393	8.343	10.656	12.242	14.684	16.919	21.666	27.877
10	2.558	3.940	4.865	6.179	7.267	9.342	11.781	13.442	15.987	18.307	23.209	29.588
11	3.053	4.575	5.578	6.989	8.148	10.341	12.899	14.631	17.275	19.675	24.725	31.264
12	3.571	5.226	6.304	7.807	9.034	11.340	14.011	15.812	18.549	21.026	26.217	32.910
13	4.107	5.892	7.042	8.634	9.926	12.340	15.119	16.985	19.812	22.362	27.688	34.528
14	4.660	6.571	7.790	9.467	10.821	13.339	16.222	18.151	21.064	23.685	29.141	36.123
15	5.229	7.261	8.547	10.307	11.721	14.339	17.322	19.311	22.307	24.996	30.578	37.697
16	5.812	7.962	9.312	11.152	12.624	15.338	18.418	20.465	23.542	26.296	32.000	39.252
17	6.408	8.672	10.085	12.002	13.531	16.338	19.511	21.615	24.769	27.587	33.409	40.790
18	7.015	9.390	10.865	12.857	14.440	17.338	20.601	22.760	25.989	28.869	34.805	42.312
19	7.633	10.117	11.651	13.716	15.352	18.338	21.689	23.900	27.204	30.144	36.191	43.820
20	8.260	10.851	12.443	14.578	16.266	19.377	22.775	25.038	28.412	31.410	37.566	45.315
21	8.897	11.501	13.240	15.445	17.182	20.377	23.858	26.171	29.615	32.671	38.932	46.797
22	9.542	12.338	14.041	16.314	18.101	21.337	24.939	27.301	30.813	33.924	40.289	48.268
23	10.196	13.091	14.848	17.187	19.021	22.337	26.018	28.429	32.007	35.172	41.638	49.728
24	10.856	13.848	15.659	18.062	19.943	23.337	27.096	29.553	33.196	36.415	42.980	51.179
25	11.524	14.611	16.473	18.940	20.867	24.337	28.172	30.675	34.382	37.652	44.314	52.620
26	12.198	15.379	17.292	19.820	21.792	25.336	29.246	31.795	35.563	38.885	45.642	54.052
27	12.879	16.151	18.114	20.703	22.719	26.336	30.319	32.912	36.741	40.113	46.963	55.476
28	13.565	16.928	18.939	21.588	23.647	27.336	31.391	34.027	37.916	41.337	48.278	56.893
29	14.256	17.708	19.768	22.475	24.577	28.336	32.461	35.139	39.087	42.557	49.588	58.302
30	14.953	18.493	20.599	23.364	25.508	29.336	33.530	36.250	40.256	43.773	50.892	59.703

APPENDIX F

PERIODIC TABLE OF

4 9.01 **Be**	
hcp	Crystal structure
2.27	*a* lattice parameter, Å
3.59	*c* lattice parameter, Å
121.	Atomic density, atoms nm^{-3}

IA	IIA	IIIB	IVB	VB	VIB	VIIB	VIII	VIII	VIII
1 1.008 **H**									
3 6.94 **Li** bcc 3.491 47.0	4 9.01 **Be** hcp 2.27 3.59 121.								
11 22.99 **Na** bcc 4.225 26.52	12 24.31 **Mg** hcp 3.21 5.21 43.0								
19 39.1 **K** bcc 5.225 14.02	20 40.08 **Ca** fcc 5.58 23.0	21 44.96 **Sc** hcp 3.31 5.27 42.7	22 47.9 **Ti** hcp 2.95 4.68 56.6	23 50.94 **V** bcc 3.03 72.1	24 52.0 **Cr** bcc 2.88 83.3	25 54.94 **Mn** cubic complex 81.8	26 55.85 **Fe** bcc 2.87 84.9	27 58.93 **Co** hcp 2.51 4.07 89.7	28 58.7 **Ni** fcc 3.52 91.4
37 85.47 **Rb** bcc 5.585 11.48	38 87.62 **Sr** fcc 6.08 17.8	39 88.91 **Y** hcp 3.65 5.73 30.2	40 91.22 **Zr** hcp 3.23 5.15 42.9	41 92.91 **Nb** bcc 3.30 55.3	42 95.94 **Mo** bcc 3.15 64.2	43 98.91 **Tc** hcp 2.74 4.40 70.4	44 101.07 **Ru** hcp 2.71 4.28 73.6	45 102.91 **Rh** fcc 3.80 72.6	46 106.4 **Pd** fcc 3.89 68.0
55 132.91 **Cs** bcc 6.045 9.05	56 137.34 **Ba** bcc 5.02 16.0	57 138.91 **La** hex. 3.77 ABAC 27.0	72 148.89 **Hf** hcp 3.19 5.05 45.2	73 180.95 **Ta** bcc 3.30 55.5	74 183.85 **W** bcc 3.16 63.1	75 186.2 **Re** hcp 2.76 4.46 68.0	76 190.20 **Os** hcp 2.74 4.32 71.4	77 192.22 **Ir** fcc 3.84 70.7	78 195.09 **Pt** fcc 3.92 66.3

MATERIAL PARAMETERS

IB	IIB	IIIA	IVA	VA	VIA	VIIA	
							2 4.003 He
		5 10.81 **B** rhomb. 130.	6 12.01 **C** diamond 3.567 176.	7 14.01 N	8 16.0 O	9 19.0 F	10 20.18 Ne 43.6
		13 26.98 **Al** fcc 4.05 60.2	14 28.09 **Si** diamond 5.430 50.0	15 30.97 **P** complex	16 32.06 **S** complex	17 35.45 Cl	18 39.95 Ar 26.6
29 63.55 **Cu** fcc 3.61 84.7	30 65.38 **Zn** hcp 2.66 4.95 65.5	31 69.72 **Ga** complex 51.0	32 72.59 **Ge** diamond 5.658 44.2	33 74.92 **As** rhomb. 46.5	34 78.96 **Se** hex. chains 36.7	35 79.9 Br 23.6	36 83.8 Kr 21.7
47 107.87 **Ag** fcc 4.09 58.5	48 112.40 **Cd** hcp 2.98 5.62 46.4	49 114.82 **In** tetr. 3.25 4.95 38.3	50 118.69 **Sn** diamond 6.49 36.2	51 121.75 **Sb** rhomb. 33.1	52 127.6 **Te** hex. chains 29.4	53 126.9 **I** complex 23.6	54 131.3 Xe 16.4
79 196.97 **Au** fcc 4.08 59.0	80 200.59 Hg 42.6	81 204.37 **Tl** hcp 3.46 5.52 35.0	82 207.19 **Pb** fcc 4.95 33.0	83 208.98 **Bi** rhomb. 28.2	84 (209) Po sc 3.34 26.7	85 (210) At	86 (222) Rn

APPENDIX G

PERIODIC TABLE OF

Key (Ni shown as an example):

- Atomic Number → 28
- Symbol → Ni
- Atomic Weight → 58.7
- Isotopes → Isotope Abundance, %: 58 68.27; 60 26.10; 61 1.13; 62 3.59; 64 0.91

IA	IIA	IIIB	IVB	VB	VIB	VIIB	VIII	VIII	VIII
1 **H** 1.008 1 99.985 2 0.015									
3 **Li** 6.94 6 7.5 7 92.5	4 **Be** 9.01 9 100								
11 **Na** 22.99 23 100	12 **Mg** 24.31 24 78.99 25 10.00 26 11.01								
19 **K** 39.1 39 93.258 40 0.011 41 6.730	20 **Ca** 40.08 40 96.941 42 0.647 43 0.135 44 2.086 46 0.004 48 0.187	21 **Sc** 44.96 45 100	22 **Ti** 47.9 46 8.0 47 7.3 48 73.8 49 5.5 50 5.4	23 **V** 50.94 50 0.25 51 99.75	24 **Cr** 52.0 50 4.345 52 83.79 53 9.50 54 2.365	25 **Mn** 54.94 55 100	26 **Fe** 55.85 54 5.9 56 91.72 57 2.1 58 0.28	27 **Co** 58.93 59 100	28 **Ni** 58.7 58 68.27 60 26.10 61 1.13 62 3.59 64 0.91
37 **Rb** 85.47 85 72.17 87 27.83	38 **Sr** 87.62 84 0.56 86 9.86 87 7.00 88 82.58	39 **Y** 88.91 89 100	40 **Zr** 91.22 90 51.45 91 11.22 92 17.15 94 17.38 96 2.80	41 **Nb** 92.91 93 100	42 **Mo** 95.94 92 14.84 94 9.25 95 15.92 96 16.68 97 9.55 98 24.13 100 9.63	43 Tc 98.91	44 **Ru** 101.07 96 5.54 98 1.86 99 12.7 100 12.6 101 17.1 102 31.6 103 18.6	45 **Rh** 102.91 103 100	46 **Pd** 106.4 102 1.02 104 11.14 105 22.33 106 27.33 108 26.46 110 11.72
55 **Cs** 132.91 133 100	56 **Ba** 137.34 130 0.106 132 0.101 134 2.42 135 6.593 136 7.85 137 11.23 138 71.70	57 **La*** 138.91 138 0.09 139 99.91	72 **Hf** 148.89 174 0.16 176 5.21 177 18.60 178 27.30 179 13.63 180 35.10	73 **Ta** 180.95 180 0.012 181 99.998	74 **W** 183.85 180 0.14 182 26.41 183 14.40 184 30.64 186 28.41	75 **Re** 186.2 185 37.4 187 62.6	76 **Os** 190.2 184 0.02 186 1.58 187 1.6 188 13.3 189 16.1 190 26.4 192 41.0	77 **Ir** 192.22 191 37.3 193 62.7	78 **Pt** 195.09 190 0.01 192 0.79 194 32.9 195 33.8 196 25.3 198 7.2
87 Fr 232.04	88 Ra (231)	89 Ac 238.03							

*						
58 **Ce** 140.12 136 0.19 138 0.25 140 88.43 142 11.13	59 **Pr** 140.91 141 100	60 **Nd** 144.24 142 27.13 143 12.18 144 23.80 145 8.30 146 17.19 148 5.76 150 5.64	61 Pm (147)	62 **Sm** 150.4 144 3.1 147 15.0 148 11.3 149 13.8 150 7.4 152 26.7 154 22.7	63 **Eu** 151.96 151 47.8 153 52.2	64 **Gd** 157.25 152 0.20 154 2.18 155 14.80 156 20.47 157 15.65 158 24.84 160 21.86

ISOTOPE ABUNDANCES

IB	IIB	IIIA	IVA	VA	VIA	VIIA	VIIIA
							2 He 4.003 3 0.00013 4 99.99987
		5 B 10.81 10 19.9 11 80.1	6 C 12.01 12 98.9 13 1.1	7 N 14.01 14 99.63 15 0.37	8 O 16.0 16 99.76 17 0.04 18 0.20	9 F 19.0 19 100	10 Ne 20.18 20 90.48 21 0.27 22 9.25
		13 Al 26.98 27 100	14 Si 28.09 28 92.2 29 4.67 30 3.10	15 P 30.97 31 100	16 S 32.06 32 95.02 33 0.75 34 4.21 36 0.02	17 Cl 35.45 35 75.77 37 24.23	18 Ar 39.95 36 0.337 38 0.63 40 99.60
29 Cu 63.55 63 69.17 65 30.83	30 Zn 65.38 64 48.6 66 27.9 67 4.1 68 18.8 70 0.6	31 Ga 69.72 69 60.108 71 39.892	32 Ge 72.59 70 20.5 72 27.4 73 7.8 74 36.5 76 7.8	33 As 74.92 75 100	34 Se 78.96 74 0.9 76 9.1 77 7.6 78 23.6 80 49.9 82 8.9	35 Br 79.9 79 50.69 81 49.31	36 Kr 83.8 78 0.35 80 2.25 82 11.6 83 11.5 84 57.0 86 17.3
47 Ag 107.87 107 51.839 109 48.161	48 Cd 112.4 106 1.25 108 0.89 110 12.49 111 12.80 112 24.13 113 12.22 114 28.73 116 7.49	49 In 114.82 113 4.3 115 95.7	50 Sn 118.69 112 0.97 114 0.65 115 0.36 116 14.53 117 7.68 118 24.22 119 8.58 120 32.59 122 4.63 124 5.79	51 Sb 121.75 121 57.4 123 42.6	52 Te 127.6 120 0.095 122 2.59 123 0.905 124 4.79 125 7.12 126 18.93 128 31.70 130 33.87	53 I 126.9 127 100	54 Xe 131.3 124 0.10 126 0.09 128 1.91 129 26.4 130 4.1 131 21.2 132 26.9 134 10.4
79 Au 196.97 197 100	80 Hg 200.59 196 0.15 198 10.0 199 16.9 200 23.1 201 13.2 202 29.8 204 6.85	81 Tl 204.37 203 29.524 205 70.476	82 Pb 207.19 204 1.4 206 24.1 207 22.1 208 52.4	83 Bi 208.98 209 100	84 Po (209)	85 At (210)	86 Rn (222) 136 8.9

65 Tb 158.93 159 100	66 Dy 162.5 156 0.06 158 0.10 160 2.34 161 18.9 162 25.5 163 24.9 164 28.2	67 Ho 164.93 165 100	68 Er 167.26 162 0.14 164 1.61 166 33.6 167 22.95 168 26.8 170 14.9	69 Tm 168.93 169 100	70 Yb 173.04 168 0.13 170 3.05 171 14.3 172 21.9 173 16.12 174 31.8 176 12.7	71 Lu 174.97 175 97.41 176 2.59

Index

MIX
Papier aus verantwortungsvollen Quellen
Paper from responsible sources
FSC® C105338

If you have any concerns about our products,
you can contact us on
ProductSafety@springernature.com

In case Publisher is established outside the EU,
the EU authorized representative is:
Springer Nature Customer Service Center GmbH
Europaplatz 3, 69115 Heidelberg, Germany

Printed by Libri Plureos GmbH
in Hamburg, Germany